Climate Change Mitigation Actions in Five Developing Countries

Five case studies on mitigation actions (MAs) in developing countries illustrate the rich diversity of climate action. Researchers from Brazil, Chile, Colombia, Peru and South Africa reflect on what is possible in their countries. These case studies reflect the sheer diversity of NAMAs: from a 'Pronami' on efficient lighting in Peru, to longer-term challenges of rising energy emissions in Brazil, and much else. The book compares the similarities and differences across eight elements that could assist in developing and implementing mitigation actions (MAs). The comparative analysis highlights how challenging implementation can be in the context of development and also points to factors that might enable ambitious mitigation. The comparison suggests that choice of MAs may be linked to institutional capacity, the resources a country is endowed with, and hence its emissions profile. International support can be an important global enabler. The authors find that addressing both development and climate objectives is key. This book fills an important gap in the literature from developing country authors about mitigation actions in their own countries.

This book was published as a special issue of *Climate and Development*.

Harald Winkler is Director of the Energy Research Centre, University of Cape Town. He co-directs the programme Mitigation Action Plans and Scenarios (MAPS), out of which this work emerged. His research interests are in energy, climate change and sustainable development.

Climate Change Mitigation Actions in Five Developing Countries

Edited by
Harald Winkler
with the assistance of Kim Coetzee

Routledge
Taylor & Francis Group
LONDON AND NEW YORK

First published 2015
by Routledge
2 Park Square, Milton Park, Abingdon, Oxfordshire OX14 4RN

and by Routledge
711 Third Avenue, New York, NY 10017, USA

First issued in paperback 2017

Routledge is an imprint of the Taylor & Francis Group, an informa business

British Library Cataloguing in Publication Data
A catalogue record for this book is available from the British Library

ISBN 13: 978-1-138-06154-5 (pbk)
ISBN 13: 978-1-138-84614-2 (hbk)

Typeset in Times New Roman
by RefineCatch Limited, Bungay, Suffolk

Publisher's Note
The publisher accepts responsibility for any inconsistencies that may have arisen during the conversion of this book from journal articles to book chapters, namely the possible inclusion of journal terminology.

Contents

Citation Information

The chapters in this book were originally published in *Climate and Development*, volume 6, issue S1 (2014). When citing this material, please use the original page numbering for each article, as follows:

Chapter 1
Introduction: Emerging lessons on designing and implementing mitigation actions in five developing countries
Harald Winkler
Climate and Development, volume 6, issue S1 (2014) pp. 1–3

Chapter 2
The international policy context for mitigation actions
Kim Coetzee and Harald Winkler
Climate and Development, volume 6, issue S1 (2014) pp. 4–11

Chapter 3
A case study on Colombian mitigation actions
Ricardo Delgado, Angela Inés Cadena, Mónica Espinosa, Catalina Peña and Mateo Salazar
Climate and Development, volume 6, issue S1 (2014) pp. 12–24

Chapter 4
Climate change mitigation actions in Brazil
Emilio Lèbre La Rovere, Amaro Olimpio Pereira Jr., Carolina Burle Schmidt Dubeux and William Wills
Climate and Development, volume 6, issue S1 (2014) pp. 25–33

Chapter 5
A case study of Chilean mitigation actions
José Eduardo Sanhueza and Felipe Andrés Ladron de Guevara
Climate and Development, volume 6, issue S1 (2014) pp. 34–42

Chapter 6
A case study of Peru's efficient lighting nationally appropriate mitigation action
Pia Zevallos, Talia Postigo Takahashi, Maria Paz Cigaran and Kim Coetzee
Climate and Development, volume 6, issue S1 (2014) pp. 43–48

Chapter 7
A case study of South African mitigation actions
Emily Tyler, Anya Sofie Boyd, Kim Coetzee and Harald Winkler
Climate and Development, volume 6, issue S1 (2014) pp. 49–58

CITATION INFORMATION

Chapter 8

Comparative analysis of five case studies: commonalities and differences in approaches to mitigation actions in five developing countries

José Alberto Garibaldi, Harald Winkler, Emilio Lèbre La Rovere, Angela Cadena, Rodrigo Palma, José Eduardo Sanhueza, Emily Tyler and Marta Torres Gunfaus

Climate and Development, volume 6, issue S1 (2014) pp. 59–70

Please direct any queries you may have about the citations to clsuk.permissions@cengage.com

INTRODUCTION

Emerging lessons on designing and implementing mitigation actions in five developing countries

Harald Winkler

Energy Research Centre and MAPS programme, University of Cape Town, UCT, Cape Town, South Africa

Introduction

This Special Issue brings together case studies on mitigation actions (MAs) by researchers in Brazil, Chile, Colombia, Peru and South Africa. The case studies illustrate the rich diversity of MAs as they are conceptualized, designed and moved towards implementation in different circumstances. The wide range of MAs enables a comparative analysis, shedding light on what is common across different cases and what is specific to each country's experience. An empirically-grounded approach to MAs emerges in this way, and is complemented by an understanding of how the concept of nationally appropriate mitigation actions (NAMAs) emerged in the international negotiations (Coetzee & Winkler, 2013).

The sheer diversity of NAMAs is a striking feature even from the relatively small set of case studies in this issue. Zevallos et al. (2014) hone in on the most advanced 'Pronami' in Peru, a case study of energy-efficient lighting. The contribution from UniAndes in Colombia considers electric vehicles (Cadena Monroy, Delgado, Espinosa, Peña, & Salazar, 2014). Other authors take a broader view, with Tyler, Boyd, Coetzee, and Winkler (2013) choosing four different MAs that might be developed in South Africa: bus rapid transit, sustainable settlements, an initiative on renewable energy and a carbon tax. Sanhueza and Ladron (2013) present a range of MAs in the transport, agriculture and energy sectors, and their development across different periods. Wills and his co-authors discuss how Brazil might meet its voluntary mitigation goals, with avoided deforestation making a major contribution in the near-term, but the longer-term challenges of low-emissions energy development being new to Brazil. The particular examples chosen by the authors in no way suggest that these are the only MAs undertaken in their respective countries; but merely that the selected MAs were found to have illustrative value.

Garibaldi et al. (2013), working with a team including authors from each of the case studies, provides a comparative analysis across the case studies. The MAs in the case studies from five countries are assessed using the following elements: conceptual approaches; the stage of development of actions; various capacities – institutional capacity for planning MAs and regulatory context, as well as technical capacity to design and measurable, reportable and verifiable (MRV) MAs; priority given to poverty; ownership of MAs; and finance.

Conceptual and methodological approach

A case study approach is a bottom-up approach to analysing how MAs are conceptualized in each country. This approach was adopted in the context of both a lack of clarity at international level regarding the definition of NAMAs and a desire to understand what was already taking place 'on the ground' in the countries. A more top-down conceptualization of NAMAs could have emerged from the international negotiations. As Coetzee and Winkler outline, however, the NAMA concept in the negotiations has thus far provided scope for diversity. Nonetheless, two of the case studies indicate that NAMAs are understood, at least as one important dimension, in the context of international pledges.

The case study from South Africa (SA) explicitly outlines a bottom-up methodological perspective in this sense: first, the South African approach to individual MAs is outlined; next, challenges to their implementation are examined; and third, possible means of overcoming such challenges – both domestic and international – are explored (Tyler et al., 2013).

In all the case studies, the importance of linking to national developmental priorities is emphasized. There are multiple and diverse developmental objectives, yet it is observed that alleviating poverty remains a high priority across all five countries. Using mitigation actions to address high levels of inequality emerges as an equally important priority in the Brazilian and SA case studies.

Sanhueza and Ladron reflect on the various periods over which understanding of MAs evolved in Chile. Early thinking followed the signing of the Convention (1992); with learning on the Clean Development Mechanism (CDM) facilitated once a prompt start was agreed in Marrakech (2001); reinforced when the Kyoto Protocol entered into force (Montreal, 2005); with a pledge developed – as for other developing countries – shortly before Copenhagen (2009). Chile communicated that it will achieve a 20% reduction below the 'business as usual' emissions growth trajectory in 2020. The current period, since 2010, has focused on how MAs can be implemented to achieve that pledge – though national circumstances remain important.

The Brazilian pledge in Copenhagen was unusual in that it contained both an overall deviation below business-as-usual (BAU) and individual NAMAs. Most other developing countries pledged either an aggregate NAMA (expressed as deviation below BAU or carbon intensity of gross domestic product) or individual NAMAs. La Rovere, Pereira, Dubeux, and Wills (2013) explain how deforestation is a major contribution to the overall deviation of 36.1% and 38.9% below BAU for 2020. They find that Brazil can meet its voluntary mitigation goals. If this was realized, a contribution by a major developing countries like Brazil would in turn provide momentum to the multi-lateral negotiations.

Challenges of implementation

The challenges of implementation for Brazil are longer-term future. La Rovere et al. provide data on the clean energy mix that Brazil has had historically, with hydro-electricity and biofuels. After 2020, Brazil will face challenges of keeping its energy mix clean, while consumption and production of fossil fuels will increase – and economic development continues to increase energy demand. Without additional MAs, Brazilian emissions would start to increase again in 2020–2030. The case study points out that this challenge is more akin to that faced by industralized countries, and indeed some other developing countries.

The policy environment in which implementation of MAs is attempted is a crucial factor. The case studies include a range of policy-making contexts, from the highly planned to the less centrally coordinated. Planning horizons differ from four years (Colombia) to 40-year scenarios (in SA, and 2050 scenarios due to be developed in other Mitigation Action Plans and Scenarios [MAPS] countries). None of these is 'better' or 'worse' – it is simply important to understand that each policy and regulatory context brings with it specific challenges.

Other challenges to implementation are illustrated in each of the case studies. Zevallos et al. explored the NAMA that they considered the most advanced in Peru. Efficient lighting is to be introduced in the residential, public, commercial and industrial sectors. The researchers found that even this NAMA was still in a 'readiness' stage at time of writing. At least one of the reasons is the complexity of governance across multiple sectors. The challenge for PlanCC (as MAPS is called in Peru) is how to move the MA beyond readiness to the design and implementation phase. It seems South Africa is at a similar stage, exhibiting strengths in identifying, analysing and designing MAs, but falling short on implementation (Tyler et al., 2013). They identify two key challenges – resistance from vested interested and the availability of finance, Which brings us to enabling factors.

Enabling factors – domestic and international

Why do countries choose particular MAs? The comparative analysis suggests that the choice of MAs may be linked to institutional capacity (both for design and implementation of MAs and possible MRV), emissions profile and the relative resource endowments of countries.

To understand specific enabling factors, however, one needs to delve into the textured accounts and detailed examples. To continue with the South African case study, Tyler et al. suggest that that a supportive policy, regulatory and planning context is an important enabling factor for implementation. More specifically, they argue that alignment of an MA with the existing priorities of a particular sector may be helpful in achieving implementation. Whether sector-specific MAs generally prove to be more effective than cross-sectoral MAs is a topic worth further investigation. Zevallos et al. point to a need for increased coordination between sectors and key stakeholders, indicating that in their view, the potential for cross-sectoral MAs is worth exploring.

Cadena Monroy et al. direct their focus to mitigation actions at the local level in Bogota. Bogota had early success with TransMillenio, but the researchers from Uni-Andes (the core research team in MAPS Colombia) focus on another part of the transport sector in their case study. They examine the potential for electric vehicles for private, utility and public services. They find that increased electricity demand does not increase greenhouse gas (GHG) emissions in such a scenario, given high proportion of hydro-electricity in the Colombian grid. A crucial developmental goal is the security of energy supply. This case study finds that diversification of the energy mix has benefits for energy security.

Sanhueza and Ladron highlight the lessons from the CDM in Chile. They find that learning in the second and third periods (2001–2009) from CDM projects provided a useful basis for the development of NAMAs.

A strong finding across all case studies is the importance of international support. Garibaldi et al.'s comparison finds that international support, particularly in the form of finance, is emphasized in all case studies. They also add that technical capacities for design and MRV of implementation exist across all five developing countries examined here.

The concept of NAMAs itself is still emergent in the international context. The need for a diversity of actions to be defined as NAMAs is emphasized in United Nations Framework Convention on Climate Change (UNFCCC) negotiations, and richly illustrated in these country case studies. The nascent nature of the concept presents an opportunity for developing countries to shape NAMAs to their benefit and take action appropriate to locally required development. The comparative analysis emphasizes the diversity of MAs and capabilities, and connected to that, a need for flexibility in definition, design, and implementation.

Increasing ambition by a focus on development

The case studies seem to indicate willingness to take action by these developing countries, and signs of putting capacities in place, albeit at different levels of maturity. If this proves to be the case, then it may be that the path to more ambitious mitigation action in these countries lies addressing both development and climate.

In conclusion, and without repeating the discussions or finding presented in the case studies, it is worth highlighting one overarching finding: MAs that are driven by both development and climate objectives are more likely to be implemented, than those driven by mitigation objectives alone.

Acknowledgements

The articles in this Special Issue have emerged from research conducted by the authors as part of the Mitigation Action Plans and Scenarios (MAPS programme). The authors, who are researchers involved in the MAPS programme, had the opportunity to share earlier versions in a MAPS Research Lab held in Lima, Peru (September 2011) and appreciated the vibrant discussion in the MAPS community subsequently. The articles were significantly reworked in preparation for submission and in response to valuable comments received from *Climate and Development's* peer reviewer process.

MAPS is a collaboration amongst developing countries to establish the evidence base for the long-term transition to robust, carbon efficient economies.

In this way, MAPS contributes to ambitious climate change mitigation that aligns economic development with poverty alleviation. MAPS current works in-country in Brazil, Chile, Colombia and Peru, with a MAPS international team based in South Africa at the Energy Research Centre (University of Cape Town) and SouthSouthNorth. We gratefully acknowledge funding by Children's Investment Fund Foundation for the original research. The responsibility for the views expressed in this guest editorial and the articles in this Special Issue remains that of the authors.

References

Cadena Monroy, Á.I., Delgado, R., Espinosa, M., Peña, C. & Salazar, M. (2014). A case study on Colombian mitigation actions. *Climate and Development.* doi:10.1080/17565529.2013.857587.

Coetzee, K., & Winkler, H. (2013). The international policy context for mitigation actions. *Climate and Development.* doi:10.1080/17565529.2013.867245.

Garibaldi, J.A., Winkler, H., La Rovere, E.L., Cadena, A., Palma, R., Sanhueza, J.E., & Torres Gunfaus, M. (2013). Comparative analysis of five case studies: Commonalities and differences in approaches to mitigation actions in five developing countries. *Climate and Development.* doi:10.1080/17565529.2013.812031.

La Rovere, E.L., Pereira, A.O., Dubeux, C.B.S., & Wills, W. (2013). Climate change mitigation actions in Brazil. *Climate and Development.* doi:10.1080/17565529.2013.812952.

Sanhueza, J.E., & Ladron de Guevara, F.A. (2013). A case study of Chilean mitigation actions. *Climate and Development.* doi:10.1080/17565529.2013.844675.

Tyler, E., Boyd, A., Coetzee, K., & Winkler, H. (2013). A case study of South African mitigation actions (For the special issue on mitigation actions in five developing countries). *Climate and Development.* doi:10.1080/17565529.2013.768175.

Zevallos, P., Takahashi, T.P., Cigaran, M.P., & Coetzee, K. (2014). A case study of Peru's efficient lighting nationally appropriate mitigation action. *Climate and Development.* doi:10.1080/17565529.2013.867251.

The international policy context for mitigation actions

Kim Coetzee and Harald Winkler

Energy Research Centre, University of Cape Town (UCT), Rondebosch, Cape Town, South Africa

In order to provide a framework for the country case studies that follow, this review article outlines how the concept of nationally appropriate mitigation actions (NAMAs) arose in the context of United Nations Framework Convention on Climate Change (UNFCCC) negotiations. The paper outlines how the NAMA concept is being developed internationally in order to juxtapose this with the country level, "bottom-up" understandings evident in the mitigation action country studies that follow. This article undertakes a review of the UNFCCC Conference of the Parties decisions from Bali in 2007 to Doha in 2012 to trace the historical emergence of NAMAs, before analysing the increasing institutionalization of NAMAs using both primary and secondary sources. The review suggests that the still-nascent nature of NAMAs may provide a vehicle for developing countries to participate in the international mitigation effort – with the technical and financial support of developed countries, subject to the Convention's principle of common but differentiated responsibilities (CBDR). The fact that there remains some lack of clarity on what constitutes a NAMA may represent an opportunity for developing countries to shape it to their benefit by providing thought-leadership and taking domestic action suitable to their developmental needs.

1. Introduction

This review article provides the international policy context for mitigation actions (MAs) so as to position the country studies of MAs that follow in this special issue within the context of internationally negotiated efforts under the United Nation Framework Convention on Climate Change (UNFCCC). To do this, the article will discuss the negotiated historical delineation between developed and developing countries in the Framework Convention and the emergence and evolution of the concept of nationally appropriate mitigation actions (NAMAs). The next section analyses the institutionalization of NAMAs before concluding with a brief discussion of how this internationally negotiated concept interacts with the domestically situated MAs of the country studies in this volume.

Many factors, including the negotiated delineation between Annex I and NAI, the changing geopolitical, economic and social dynamics between developed and developing countries, the increasing need for urgent action as evident from climate science and observed impacts, and the concept of equitable burden sharing, all underpin the importance of the current international context of NAMAs. Specifically, the primary importance of NAMAs is as the main vehicle for supporting – through technology, financing and capacity-building – the voluntary developing country mitigation efforts. Efforts to mitigate emissions (by deviating from projected 2020 baseline emissions) are also required by developing countries in order to achieve low to medium greenhouse gas (GHG) concentration levels (450 ppm CO_2eq) – actions by developed countries are necessary but insufficient (IPCC, 2007, pp. 775–776).[1]

2. Historical development of NAMAs in the UNFCCC

The negotiations between sovereign states on limiting GHG emissions gained a formal institution with the signing of the UNFCCC in 1992. As is common in the practise of global environmental governance, the signing of a framework agreement or convention between sovereign states is a first step indicating an intention to address an issue without committing to specific actions (DeSombre, 2007). Thus, the UNFCCC (1995) contained no legally binding quantified commitments to emissions reductions, but in 1995 at COP1 in Berlin, negotiations were initiated on "a protocol or another legal instrument" that would revise the inadequate commitments of the industrialized countries under the Convention. These "Berlin Mandate" negotiations eventually lead to the adoption of the Kyoto Protocol (UNFCCC, 1998) at COP3 in 1997 in Japan (Yamin & Depledge, 2004).

The UNFCCC does, however, include four key elements. First, it outlines a long-term environmental objective of "stabilising GHG concentrations in the atmosphere at a level that would prevent dangerous anthropogenic interference with the climate system" (United Nations, 1992, Article 2). Second, a clear principle that the balance of responsibilities of developed and developing countries should be distributed "in accordance with their common but differentiated responsibilities and respective capabilities" (CBDR & RC) (United Nations, 1992, Article 3). This principle is also evident in the differentiated nature of the commitments in Article 4 and in the acknowledgement in Article 4.7, that ability of developing countries to implement their responsibilities would be contingent on the implementation of commitments by developed countries (Winkler & Rajamani, 2013). Third, a range of commitments: those applicable to both developed and developing countries are outlined qualitatively in Article 4.1. whereas additional short-term emissions stabilization goals applicable only to Annex I countries are outlines under Article 4.2, (a) and (b) ; United Nations, 1992). Fourth, the means of implementing those commitments: financial support (Article 4.3 and 4.4), technology transfer (Article 4.5). Signatories to the Kyoto Protocol undertook quantified emission limitation or reduction objectives (QELROs), and were entitled to use market mechanisms (Emissions Trading, the Clean Development Mechanism and Joint Implementation) in order to meet those commitments.

In the context of these four key elements however, the debate over the need for and form of MAs by developing countries has a history almost as long as the multilateral climate change negotiations themselves. In the intervening 20 years since the UNFCCC was signed, the global geopolitical and economic landscape has altered significantly, and this has been reflected in calls for a re-framing of the terms of the attribution of differentiated responsibilities and corresponding need for action. One aspect that has remained remarkably consistent since at least 1997s "Byrd–Hagel"[2] Senate Resolution, however, has been the USA's insistence it will not take on any quantified reduction commitments in the absence of the same for large developing countries, most notably China.

In the opening years of the new millennium, the GHG emissions of the larger developing countries – like China – began to increase steadily as their economies grew after the slump of the late 1990s. However, addressing poverty and encouraging development remained a prime focus of policy-makers in many developing countries, with GHG emissions some way down the list of priorities. An early attempt to balance both development needs and mitigation of GHG emissions was the concept of Sustainable Development Policy and Measures, or SD-PAMS. In conceiving an SD-PAM, developing country decision-makers would begin by identifying the development priorities and only then isolate the policies and measures that provided synergies between sustainable development and reductions of GHG emissions (Winkler, Spalding-Fecher, Mwakasonda, & Davidson, 2002). At COP 8 in 2002, the Delhi Ministerial Declaration on Climate Change and Sustainable Development (UNFCCC, 2003, 1/CP.8) outlined the importance of integrating climate change policies and measures with national development programmes in a manner that took into account "their specific national and regional development priorities, objectives and circumstances" – this language anticipates the use of "nationally appropriate" in relation to MAs. The idea of SD-PAMs gained traction in the mid-2000s – it was incorporated into the South African Government submission to the UNFCCC's Convention Dialogue held during 2006 (RSA, 2006) and featured in presentations by the Annex I Expert Group on the UNFCCC (Ellis, Baron, & Buchner, 2007) at the conference of the parties (COP) in 2007. During the Bali COP negotiations, however, the SD-PAMs terminology was dropped in favour of the term "NAMAs".

2.1. *Bali action plan*

The words "NAMAs" in relation to developing countries appeared for the first time in paragraph 1(b) (ii) of the Bali Action Plan (BAP; UNFCCC, 2008). The text of this COP decision in paragraph 1(b) (i) also calls on developed country parties to adopt "nationally appropriate mitigation commitments or actions". The BAP introduced the concept of "measurable, reportable and verifiable" (MRV) commitments (to reduce emissions) or actions by developed countries. Article 2(b)ii states that MRV is only of the *actions* of developing countries; actions that would be undertaken with support provided by developed countries in the form of finance, technology and capacity-building (UNFCCC, 2008). In Bali, developing countries agreed to make their actions quantifiable (MRV) subject to support, while all developed countries (including the USA that was not part of Kyoto) would have the MRV of commitments or actions, including QELROs, to ensure comparable effort among AI Parties under the Convention and Protocol.

In addition to introducing the NAMA term, the BAP also established a subsidiary body under the Convention known as the Ad Hoc Working Group on Long-term Cooperative Action (AWG-LCA) which was tasked with launching, "a comprehensive process to enable the full, effective and sustained implementation of the Convention through long-term cooperative action, now, up to and beyond 2012, in order to reach an agreed outcome and adopt a decision at its fifteenth session" (UNFCCC, 2008, 1/CP13, paragraphs 1 and 2). Put simply, the BAP negotiations were intended to produce an agreement to be adopted at Copenhagen in 2009 that would come into force after the first commitment period of the KP ended in 2012. The agreement should include a shared vision, mitigation and adaptation measures, technology development and transfer, and investment and financial assistance (Gupta, 2010). Despite the extra inter-sessional meetings held during 2009 the AWG-LCA was still unable to produce an agreed outcome for the COP15 to ratify. Instead it produced only a highly bracketed draft negotiating text for further negotiation at the COP in 2009 in Copenhagen. The two weeks of the COP proved to be insufficient time to resolve the outstanding issues and the end result was neither a second commitment period for the Kyoto Protocol, nor a new treaty under the Convention that could surmount the geopolitical hurdle created by the self-imposed exclusion of the USA.

2.2. *Rise of the developing countries*

COP15 also provided an overt demonstration of the continuing reorientation of geopolitics affecting the Climate Change negotiations. The 11th hour, side-room negotiations that produced the Copenhagen Accord featured the leaders of Brazil, India, South Africa and China (so-called BASIC countries) and the USA, but not the EU[3]. The Accord was not formally adopted as a COP decision and was only noted (UNFCCC, 2010, 2/CP.15) due to a lack of consensus in the plenary vote (Bodansky, 2011).

As at writing (August 2013), 141 countries have associated with the Copenhagen Accord[4]. Fifteen developed countries and the EU have provided quantified economy-wide emissions targets for 2020 which are listed in Appendix I (contained in document FCCC/SB/2011/INF.1); and 45 developing countries have provided a list of NAMAs which are listed in the Info Note (FCCC/AWGLCA/2011/INF.1). In Cancun Mexico the following year, paragraphs 48–67 of decision 1/CP16 (UNFCCC, 2011) began the process of institutionalizing NAMAs by formally integrating the Copenhagen Accord pledges and requesting the establishment of the registry. This institutionalization is analysed in depth in Section 3 of this review article.

With the conclusion in Doha of the work of the AWG-LCA (UNFCCC, 2013a, 1/CP18), some analysts have interpreted the on-going negotiations under the Durban Platform on Enhanced Action (to be concluded by 2015) as an opportunity to establish a more reciprocity-based "protocol, another legal instrument or an agreed outcome with legal force under the Convention [that is] applicable to all Parties" (UNFCCC, 2012, 1/CP17, paragraph 7) that removes the rigid division between Annex I and Non-Annex I country parties (Boyle & Aguilar, 2012). The world at the beginning of the second commitment period of the Kyoto Protocol – as per the Doha Amendment of 2012 – is self-evidently not the same as that which negotiated and then entered the first commitment period (OECD, 2010). Whilst more reciprocal regimes have emerged in other international contexts, and "against this background, the UNFCCC...of 1992 is an anachronism" (Saran, 2010) it remains true that developed countries still have the capacity to do more than developing countries. Thus whilst changes in the role of developing countries in relation to MAs may reflect broader geopolitical changes, neither these broader changes nor the Durban Platform's phrase "applicable to all Parties", should be construed as an excuse to ignore the development challenges faced by developing countries or to disregard the "UNFCCC and Kyoto models of differentiation of commitments between developed and developing country parties" (Winkler & Rajamani, 2013).

2.3. *Use beyond the UNFCCC*

The NAMA term is specifically linked to the UNFCCC negotiation process and a survey of primary and secondary literature has not revealed much attempt to use or define it beyond its use in the broader field of climate governance. Even within the broader field, however, it is often unelaborated or undefined; for instance a statement by the Major Economies Forum on Energy and Climate[5] in July 2009 noted only that all participating MEF member countries – developed and developing – would undertake NAMAs [6]. Inter-governmental partnerships such as the "International Partnership on Mitigation and MRV"[7] for instance are supported by research organizations and think tanks in the field which have been writing about NAMAs since 2007. Much of the attempt to define or clarify the concept publically has come from civil society, research, academic and NGO groups.

3. Analysis of the institutionalization of NAMAs

Paragraph 5(a) of the Copenhagen Accord records the political agreement that non-Annex I Parties to the Convention will submit their MAs on a defined form (Annex II) to the UNFCCC secretariat by 31 January 2010. In addition, the paragraph states that developing country actions will be subject to domestic measurement, reporting and verification (MRV). Three years after the words " NAMAs" appear in the BAP, COP16:

> [a]grees that developing country Parties will take nationally appropriate mitigation actions in the context of sustainable development, supported and enabled by technology, financing and capacity-building, aimed at achieving a deviation in emissions relative to 'business as usual' emissions in 2020" in paragraph 48. (UNFCCC, 2010)

As will be discussed below, the problem of defining and identifying a NAMA brings with it a new set of challenges for developing countries and the multilateral process.

3.1. *Definition, or lack thereof*

The complexity of the issue and the slow nature of negotiations have thus far resulted in a lack of conceptual clarity over precisely what constitutes a NAMA. The literature attempting to define NAMAs – written by practitioners and academics (Roser & de Vit, 2012; Torres, Winkler, Tyler, Coetzee, & Boyd, 2012; Tyler, Boyd, Coetzee, Torres Gunfaus, & Winkler, 2013; Van Tilburg, Cameron, Würtenberger, & Bakker 2011) – has been growing steadily, even though the UNFCCC only launched the prototype web-based registry for NAMAs seeking international support in April 2013[8] (UNFCCC, 2013b). Despite this, and perhaps even because of the lack of conceptual clarity, "there exists a tremendous scope for flexibility, customization and innovation" (CCAP, 2012). It was as a first step towards understanding where and how such customization and innovation might take place at a domestic level, that the country studies that follow in the special edition were conceived and undertaken. Certainly action will have to reduce emissions, at least relative to growth or a baseline scenario, to be called MAs. On the other hand, developing countries are not committing to a mitigation outcome, but rather to the implementation of the *action*. Also, the priorities of poverty and development at the national level are key in shaping how developing countries approach MAs, as higher policy priority is likely to be given to MAs that can show that they may alleviate poverty, reduce inequality or contribute to socio-economic development (Wlokas et al., 2012; Tyler, Boyd, Coetzee, & Winkler, Wlokas et al., 2013).

3.2. *What makes a NAMA "nationally appropriate"?*

Analysis of the Copenhagen Accord Appendix II submissions from the developing countries indicates the extent to which these countries see potential MAs as facilitating a broader agenda of sustainable development. The submissions encompass a range of actions from the macro-level of building or upgrading energy infrastructure (hydroelectric dams, wind farms, solar PV), to the micro-level of promoting the use of improved stoves in households; from the rural focus of reforestation and afforestation projects, low-tillage agriculture and seed preservation to more urban issues of public transport improvements and waste management (composting, recycling, co-generation). In this sense, it is clear that for many developing countries, climate change represents simply another type of development challenge – following on from debt crises, structural adjustment programs, contagious disease epidemics and HIV/AIDS for instance.

It is also clear from country studies that follow in this volume that the MAs forwarded, whilst diverse in scale and sector, are aligned with either national developmental priorities or poverty alleviation at some level. While "national circumstances" may be a mantra in the negotiations, it is also a reality that shapes developing countries' approaches. Generally, the context of poverty in developing countries makes them look at mitigation from a starting point of development (Winkler et al., 2002), and also of making that development more sustainable (Munasinghe, 2007). Specific factors play a role – so for instance the energy resources with which a country is endowed will shape which MAs it considers nationally appropriate, as might the number of heating or cooling days required for comfortable living (Baumert, Herzog, & Pershing, 2005).

3.3. *Diversity of NAMAs*

The diversity of NAMAs highlights the difficulty of the negotiations on what should and should not be conceptualized as a NAMA requiring international support. Whilst there is not yet (at time of writing August 2013) a negotiated UNFCCC decision specifically categorizing NAMAs, a frequent categorization in the literature is as follows: Emissions targets (either in terms of climate neutrality; below business as usual; below a base year and emissions intensities); strategies and plans; policies and programmes; and projects (Röser, Van Tilburg, Davis, & Höhne, 2011) Other analyses only disaggregate between NAMAs that reduce emissions and "other" – a list of areas in which action might be taken without quantified sectoral actions (Agyemang-Bonsu, 2011) and the UNEP Risø Centre's "NAMA-pipeline" categorizes NAMAs according to submission for support for preparation or implementation and submission for recognition (not requesting support) as per the UNFCCC prototype registry (UNEP Risø Centre, 2013). Still other commentators have pointed out that many actions proposed are primarily aimed at socio-economic and poverty eradication objectives rather than climate change mitigation (Van Asselt, Berseus, Gupta, & Haug, 2010). Indeed the sheer diversity of actions submitted by developing countries to the UNFCCC vividly emphasizes the recurring point that developing countries are necessarily concerned with juggling environmental, social and economic goals that any a one-size-fits-all approach would be unhelpful.

Thus far, it is still too early in the NAMA-related negotiations to be able to see any strong causality between the institutional elements analysed in Section 3 and the shape of NAMAs being developed by developing countries. For instance no NAMAs on the prototype registry have yet been "matched" with funders, which is the most likely point at which shaping or constraining might occur. It also remains to be seen whether the Subsidiary Body for Implementation (SBI) work programmes mandated to facilitate "the preparation and implementation" of NAMAs (UNFCCC, 2013a, 2/CP18, paragraph 19) through "focused interactive technical discussions … in-session workshops with input from experts and submissions from Parties and observer organizations" (UNFCCC, 2013a, 2/CP18, paragraph 20) could be an avenue for enabling or constraining diversity.

3.4. *Support for NAMAs*

The call for developing country actions which are "supported and enabled" in BAP paragraph 1(b)(ii) echoes the link created between support (by developed countries) and the extent of action by developing countries – in Article 4.7 of the Convention. Many developing countries emphasize this point in their Copenhagen Accord Annex II submissions.

The Copenhagen Accord included a pledge of US$30 billion of the so-called fast-start finance to be accessed by developing countries between 2010 and 2012 and balanced between adaptation and mitigation; and a longer term collective "goal" of mobilizing US$100 billion per year by 2020 from public and private, bilateral and multilateral sources. The Green Climate Fund to disburse these funds was officially established (1/CP16 paragraph 102) by the Cancun agreement (UNFCCC, 2010) and the governing instrument of the fund was approved in Durban (UNFCCC, 2012, 3/CP17, paragraph 2).

At COP17 it was reiterated (UNFCCC, 2012, 2/CP17, paragraph 57) that the level of implementation of voluntary mitigation commitments by developing countries depends on the degree to which developed country parties have followed through on their technology transfer and financing commitments under the Convention. This in turn links the discussion of support to the discussion of MRV of both support and actions and the registry which connects them. Developing countries continue to assert the voluntary nature of NAMAs even as some commentators have noted the increasing trend toward parity or "parallelism" between developed and developing country mitigation requirements which effectively "levels down" ambition (Rajamani, 2011).

3.5. *The NAMA registry*

The idea of a NAMA registry was first proposed by both South Africa and South Korea in submissions to the AWGLCA in 2008 and following negotiations it was finally established in principle at COP16 in Cancun (UNFCCC, 2011). The function of the NAMA registry is to record "NAMAs seeking international support and to facilitate matching of finance, technology and capacity-building support for these actions" (1/CP16, paragraph 53). The registry is tasked to keep a record of the voluntary developing country actions requiring support (including estimates of cost, emissions reductions and implementation time as these are provided); availability of developed country offers of support; and actions which have been supported (UNFCCC, 2011, 1/CP16, paragraphs 53–56). NAMAs not seeking international support are to be recorded in a separate section of the registry (UNFCCC, 2011, 1/CP16, paragraph 58).

At COP17 in Durban substance was added to proposals on the NAMA registry to reflect the complex and varied nature of the possible actions incorporated under this rubric (UNFCCC, 2012). Accordingly the registry is to be developed as a dynamic, web-based platform, with a dedicated team that populates the registry with information both explicitly and voluntarily supplied for the registry which would be structured so as to reflect the "full range of the diversity of NAMAs, and a range of types of support" (UNFCCC, 2012, 2/CP17, paragraph 45). In paragraphs 46 and 48 more details have been added in terms of information which might be provided by both recipients and donors (countries, multilateral funds and any other public, private or NGO funds) to the registry.

The UNFCCC Secretariat unveiled the prototype registry in April 2013. NAMAs can only be submitted through a National Focal Point – of which there is one per developing country – and at this stage (August 2013) the submitted NAMAs are only available to view in PDF format [9]. At this point (August 2013), the prototype NAMA registry has separated the NAMAs submitted by developing countries into three categories namely; NAMAs seeking international support for preparation; support for implementation and other NAMAs to be recognized but not requiring international support (UNFCCC, 2012, 2/CP.17, paragraphs 46 and 47). The prototype is intended as a trial run for parties to use to identify issues prior to the launch of the final registry at least two months before the 19th session of the COP in Warsaw in December 2013 (UNFCCC, 2013b, Decision 16/CP.18, paragraph 10).

The establishment of the NAMA registry is an important step to facilitate the matching of voluntary, developing country NAMAs with the "finance, technology and capacity-building support" they might need in order to be implemented (UNFCCC, 2011, 1/CP16, paragraph 53). The acknowledgement of the need for support refers back to the provision of support paragraph in the Convention (Article 4, paragraph 3) and also to the voluntary nature of the actions that take place in the context of sustainable development and poverty eradication and the greater capacity of developed country parties to provide assistance.

3.6. *Reporting*

In the negotiations since Bali, NAMAs have increasingly become associated with developing country MAs, while developed countries under the Convention commit to quantified economy-wide emission reduction targets (QEERTs; UNFCCC, 2010). While both developing and developed countries agreed to actions that will be subject to MRV, distinctions in form and content remain. In form, developed countries commit to targets that specify the mitigation outcome, while developing countries commit to implementing action (and not explicitly, at least internationally, to the reductions). In content, developed country QEERTs should reduce emissions in absolute terms compared to 1990, while developing country MA's should be designed to achieve a "deviation in emissions relative to 'business as usual' emissions in 2020" (UNFCCC, 2011, 1/CP16, paragraph 48).

Prior to the Cancun Agreement (UNFCCC, 2011, 1/CP16), non-Annex 1 countries were not obliged to submit National Communications at specified intervals (Yamin & Depledge, 2004). This has been significantly altered: developing countries must now submit National Communications to the COP every four years (1/CP16, paragraph 60(b)), as well as biennial updates (1/CP16, paragraph 60(c)) to the Subsidiary Body on Implementation (SBI) for "international consultations and analysis" (1/CP16, paragraph 63). These biennial updates should include GHG inventory information (as usually contained in National Communications) as well as extensive information on MAs e.g. impact analysis, methodologies, assumptions, costs and domestic MRV (1/CP16, paragraph 64). Thus, it is apparent that the introduction of the NAMA concept has increased the frequency and depth of the reporting requirements for developing countries[10] thereby resulting in a "levelling up" or increased symmetry between reporting requirements placed on developed and developing countries (Rajamani, 2011). In relation to NAMAs, the Durban COP adopted the guidelines for the preparation of biennial update reports by non-Annex I Parties (UNFCCC, 2012, paragraph 39), set a date for the first biennial report to be submitted – no later than December 2014[11] – and outlined modalities in relation thereto (UNFCCC, 2012, paragraph 41).

Bearing in mind that developing countries agreed to undertake NAMAs if "supported and enabled by technology, financing and capacity-building" (1/CP16, paragraph 48), an important unresolved issue with regard to MRV is that of the MRV of the provision of support. Specifically, issuing from 1/CP16, paragraph 66, COP16 left unresolved two critical points for in relation to NAMAs, namely; "measurement, reporting and verification of supported actions and corresponding support" and "domestic verification of MAs undertaken with domestic resources" (UNFCCC, 2011). The Standing Committee on Finance also has a mandate for "improving coherence and coordination" of, amongst other things, "measurement, reporting and verification of support provided to developing countries" (1/CP16, paragraph 112). This is an important consideration, given that some commentators have flagged the separation of the MRV of support and the MRV of actions – first in the Copenhagen Accord (points 4 and 5) and then in Cancun Agreements (1/CP16, paragraphs 49 and 52) – as potentially problematic for developing countries as it undermines the contingent link between the provision of the support and the eventual action (Rajamani, 2011).

4. Concluding comments

This review article has sought to provide the international policy context – the so-called "top-down approach" to policy-making – for the five developing country MAs studies – the "bottom-up approach" – that follow in this special issue. Therefore, it has traced the history of how the discussion of NAMAs by developing countries has evolved through successive UNFCCC Conferences of the Parties (COPs) and to some very limited extent in other fora.

The review concludes that NAMAs are still in the early stages, with little agreement on precise definitions. Nevertheless, the institutionalization of NAMAs continues: progress has been made in clarifying the content and role of the registry over successive COPs which is a significant step in linking support and actions. The international context will require increased measurement, reporting and verification of actions, but also offers opportunities for support in the form of finance, technology and capacity-building if developing countries can ensure that MRV of support remains on the agenda too.

Leaving aside the "nationally appropriate" part of NAMAs – which often raises associations with the international dimensions of finance and/or reporting – the range of MAs, whether in need of international support or not, shows great diversity. To a significant extent, different domestic (national or even sub-national) contexts may shape how MAs are domestically conceptualized and conceived: political cultures and particular energy resources and resource endowments typically play a part in shaping each country's approach to MAs. Even when approaching the international context from these different vantage points, a theme that emerges is the dual and sometimes competing consideration of development needs and mitigation of climate emissions.

The very multiplicity of national situations has resulted in the emergence of a wide range of MAs. It is with this in mind that the developing country studies that follow analyse select domestic i.e. "bottom-up" MAs and differentiate these from more "top-down" NAMAs, which are conceptually linked with the UNFCCC negotiations, its registry and support mechanisms. It is hoped that the analysis of the "bottom-up" MAs in this volume might lead to some illumination of both the "MAs" and the "nationally appropriate" aspects of the NAMA term. The shear diversity of NAMAs recorded in Annex II of the Copenhagen

Accord and of the country MAs further elaborated upon in this volume, highlights that stakeholders are grappling with this still ill-defined but potentially useful vehicle. It remains to be seen, however, whether the increasing institutionalization of NAMAs detracts from the diversity and hence the potential usefulness of NAMAs from the point of view of developing countries.

It is in this context that the focus of this special issue is on MAs by five developing countries: Brazil, Chile, Colombia, Peru and South Africa.

Notes

1. A 2008 article by two of the WGIII IPCC authors estimated that a deviation of between 15% and 30% below 2020 BAU scenarios would likely be required of NAI countries as a group (den Elzen & Höhne, 2008). AR5 is to be released in 2014.
2. United States of America, S.RES. 98, 105th Cong. (1997). Accessed at http://www.nationalcenter.org/KyotoSenate.html.
3. http://www.bbc.co.uk/blogs/worldtonight/2009/12/copenhagen_the_dawn_of_a_new_p.html | http://www.indianexpress.com/news/24-hours-too-late-on-telangana/560530/0| http://www.thedailybeast.com/newsweek/blogs/the-gaggle/2009/12/18/obama-dramatically-interrupts-meeting-negotiators-reach-final-agreement.html.
4. http://unfccc.int/meetings/copenhagen_dec_2009/items/5262.php [accessed 29 August 2013].
5. The MEF is composed of Australia, Brazil, Canada, China, the European Union, France, Germany, India, Indonesia, Italy, Japan, the Republic of Korea, Mexico, Russia, South Africa, the United Kingdom, and the USA.
6. http://www.majoreconomiesforum.org/past-meetings/the-first-leaders-meeting.html.
7. http://www.mitigationpartnership.net.
8. http://unfccc.int/cooperation_support/nama/items/7476.php.
9. http://unfccc.int/cooperation_support/nama/items/6945.php.
10. Reporting requirements are more flexible for Least Developed Country Parties and Small Island Developing States.
11. Least Developed Country Parties and Small Island Developing States may submit biennial update reports at their discretion and are not bound by the December 2014 deadline.

References

Agyemang-Bonsu, W.K. (2011). *Operationalization of NAMA concept*. Seoul: UNFCCC Secretariat, Financial and Technical Support Programme.

Baumert, K.A., Herzog, T., & Pershing, J. (2005). *Navigating the numbers: Greenhouse gas data and international climate policy*. Washington, DC: World Resources Institute.

Bodansky, D. (2011). The copenhagen climate change conference: A postmortem. *The American Journal of International Law, 104*(2), 230–240.

Boyle, J., & Aguilar, S. (2012). *Scenarios and sticking points under the Durban platform: The long and winding road to 2020*. London: The International Institute for Sustainable Development.

CCAP. (2012). *Overview of NAMA financial mechanisms*. CCAP. Retrieved from http://www.ccap.org/docs/resources/1135/CCAP%20NAMAs%20and%20Financial%20Mechanisms%20final.pdf

DeSombre, E.R. (2007). *The global environment and world politics*. London: Continuum.

Ellis, J. (OECD), Baron, R. (IEA), & Buchner, B. (IEA). (2007). *SD-PAMs: What, when, where, and how?* Presentation by the Annex I Expert Group on the UNFCCC at COP.

den Elzen, M., & Höhne, N. (2008). Reductions of greenhouse gas emissions in Annex I and non-Annex I countries for meeting concentration stabilisation targets. An editorial comment. *Climatic Change, 91*(3/4), 249–274.

Gupta, J. (2010, September/October). History of international climate change policy. *WIREs Climate Change*, pp. 636–653.

IPCC (Intergovernmental Panel on Climate Change). (2007). *Climate Change 2007: Mitigation. Contribution of Working Group III to the Fourth Assessment Report of the Intergovernmental Panel on Climate Change* [B. Metz, O. R. Davidson, P.R. Bosch, R. Dave, L.A. Meyer (eds)], Cambridge University Press, Cambridge, United Kingdom and New York, NY, USA.

Munasinghe, M. (2007). *Making develpoment more sustainable: Sustainomics framework and practical applications*. Colombo: MIND Press.

OECD. (2010). *Perspectives on global development: Shifting wealth*. Paris: OECD Publishing.

Rajamani, L. (2011, April). The cancun climate agreements: Reading the text, the subtext and the tea-leaves. *International & Comparative Law Quarterly, 60*, 499–519.

Röser, F., Van Tilburg, X., Davis, S., & Höhne, N. (2011). *Annual status report on nationally appropriate mitigation actions*. Ecofys, ECN and CCCP Report. Retrieved April 9, 2012, from http://www.ecofys.com/files/files/namas_annualstatusreport_2011.pdf

Röser, F., & de Vit, C. (2012). *Nationally appropriate mitigation actions (NAMAs) and carbon markets*. Policy update, Ecofys.

RSA. (2006). *Dialogue working paper 18: Submission from South Africa: Sustainable development policies and measures*. Pretoria: Department of Environmental Affairs & Tourism: 3.

Saran, S. (2010). Irresistible forces and immovable objects: A debate on contemporary climate politics. *Climate Policy, 10* (6), 678–683. Retrieved from http://www.indiaenvironmentportal.org.in/files/climatepolicy.pdf

Torres, M., Winkler, H., Tyler, E., Coetzee, K., & Boyd, A. (2012). *Mitigation action, NAMAs and low carbon development strategies: Building a common understanding of terms within MAPS*. Memo for MAPS (Mitigation Action Plans and Scenarios). Energy Research Centre, University of Cape Town. Retrieved from http://www.mapsprogramme.org/wp-content/uploads/MA-Common-Language_Memo.pdf

Tyler, E., Boyd, A., Coetzee, K., Torres Gunfaus, M., & Winkler, H. (2013). Developing country perspectives on "mitigation actions", "NAMAs", and "LCDS", *Climate Policy, 13*(6), 770–776.

Tyler, E., Boyd, A.S., Coetzee, K., & Winkler, H. (2013). A case study of South African mitigation actions (For the special issue on mitigation actions in five developing countries). *Climate and Development*. doi:10.1080/17565529.2013.768175

UNEP Risø Centre. (2013). *NAMA Pipeline*, Jørgen Fenhann (Ed). Updated 1 July 2013. Retrieved from www.namapipeline.org

UNFCCC. (1995). *Decision 1/CP.1. (FCCC/CP/1995/7/Add.1)*. Berlin: United Nations Framework Convention on Climate Change Secretariat.

UNFCCC. (1998). *Decision 1/CP.3. (FCCC/CP/1997/7/Add.1).* Kyoto: United Nations Framework Convention on Climate Change Secretariat.

UNFCCC. (2003). *Decision 1/CP.8 Delhi ministerial declaration on climate change and sustainable development (FCCC/CP/2002/7/Add.1).* New Delhi: United Nations Framework Convention on Climate Change Secretariat.

UNFCCC. (2008). *Decision 1/CP.13. Bali action plan (FCCC/CP/2007/6/Add.1).* Bali: United Nations Framework Convention on Climate Change Secretariat.

UNFCCC. (2010). *Decision 2/CP.15: Copenhagen Accord. Noted in COP Report* (FCCC/CP/2009/11/Add.1). New York: United Nations.

UNFCCC. (2011). *Decision 1/CP.16: The cancun agreements: Outcome of the work of the Ad Hoc working group on long-term cooperative action under the convention* (FCCC/CP/2010/7/Add.1). New York: United Nations.

UNFCCC. (2012). *Decision 1/CP.17: Establishment of an ad hoc working group on the Durban platform for enhanced action* (FCCC/CP/2011/9/Add.1). Durban: United Nations Framework Convention on Climate Change Secretariat.

UNFCCC. (2013a). *Decision 1/CP.18: Agreed outcome pursuant to the Bali action plan and decision 2/CP.18: Advancing the Durban platform* (FCCC/CP/2012/8/Add.1). Doha, Qatar: United Nations Framework Convention on Climate Change Secretariat.

UNFCCC. (2013b). *Decision 16/CP.18: prototype of the registry* (FCCC /CP/2012/8/Add.2). Doha: United Nations Framework Convention on Climate Change Secretariat.

United Nations. (1992). *United nations framework convention on climate change.* Rio de Janeiro: United Nations.

Van Asselt, H., Berseus, J., Gupta, J., & Haug, C. (2010). *Nationally Appropriate Mitigation Actions (NAMAs) in developing countries: Challenges and opportunities.* Report 500102 035 for the Netherlands Programme on Scientific Assessment and Policy Analysis. Amsterdam: Institute for Environmental Studies, Vrie Universiteit.

Van Tilburg, X., Cameron, L.R., Würtenberger, L., & Bakker, S.J. A. (2011). *On developing a NAMA proposal.* Discussion paper, Energy Research Centre of the Netherlands. Retrieved from http://mitigationpartnership.net/sites/default/files/ecn__nama_discussion_paper.pdf

Winkler, H., & Rajamani, L. (2013). CBDR&RC in a regime applicable to all. *Climate Policy.* doi:10.1080/14693062.2013.791184

Winkler, H., Spalding-Fecher, R., Mwakasonda, S., & Davidson, O. (2002). Sustainable development policies and measures: Starting from development to tackle climate change. In K. Baumert, O. Blanchard, S. Llosa, & J. F. Perkaus (Eds.), *Building on the Kyoto Protocol: Options for protecting the climate* (pp. 61–87). Washington, DC: World Resources Institute.

Wlokas, H.L., Rennkamp, B., Torres Gunfaus, M., Winkler, H., Boyd, A., Tyler, E., et al. (2012). *Low Carbon Development and Poverty: Exploring poverty alleviating mitigation action in developing countries.* Research Report for MAPS (Mitigation Action Plans and Scenarios). Energy Research Centre, University of Cape Town. Retrieved from http://www.mapsprogramme.org

Yamin, F., & Depledge, J. (2004). *The international climate change regime. A guide to rules, institutions and procedures.* Cambridge: Cambridge University Press.

A case study on Colombian mitigation actions

Ricardo Delgado[a], Angela Inés Cadena[a], Mónica Espinosa[a], Catalina Peña[a] and Mateo Salazar[b]

[a]*School of Engineering, Universidad de los Andes, Bogotá, Colombia;* [b]*CEDE, Universidad de los Andes, Bogotá, Colombia*

This paper introduces the Colombian case study developed under the Mitigation Actions Plans and Scenarios Initiative and the Endesa Colombia Electric Vehicles Initiative. It is aimed to consider the use of electricity in the transportation sector, particularly in the city of Bogotá, Colombia. A model for assessing the economic and environmental impacts of the introduction of electric vehicules for private, utility and public services is proposed. Diversification of the energy basket by introducing electricity in the transportation sector is a plausible development strategy for the country. Colombian electricity is relatively clean because of the high hydro component in the generation basket; therefore the substitution of fossil fuels for electricity represents net savings in emissions of greenhouse gases as well as a reduction of particulate matter. The increase in electricity demand as a result of this new use is not high enough to stress installed generation capacity or supply security.

1. Introduction

Historically, developing countries have been responsible for only a small share of the world total greenhouse gas (GHG) emissions (Baumert, Herzog, & Pershing, 2005). In contrast, in the coming years, emissions are expected to grow faster in developing countries (DOE & EIA, 2011). In this context arises one of the most complicated challenges of global climate policy: to follow low-carbon-development paths. On the one hand, developing countries are trying to reach economic prosperity as fast as possible; on the other hand, costs and availability of clean technologies may slow down the economic growth, leaving as cheaper and available alternative the use of traditional dirty technologies (Cadena et al., 2011).

As response, the international community has been working on mechanisms to support the abatement of GHG emissions in developing countries. The implementation of Nationally Appropriate Mitigation Actions (NAMAs) in the context of sustainable development is one of those mechanisms (UNFCC, 2007). 'Nationally appropriate' entails that its implementation contributes to the development goals stated by local policy. Those measures might be implemented by a developing country supported by the international community, considering that countries have common but differentiated responsibilities regarding climate change (United Nations, 1992).

The issue of climate change is on the Colombian policy agenda. The country joined the United Nations Framework Convention on Climate Change (UNFCCC) in 1994 and signed the Kyoto protocol. More recently, and following the guidelines of the National Development Plan (NDP) 2010–2014 (NPD, 2011), the Council of Economic and Social Policy (CONPES) presented an official policy report to establish the institutional arrangement that facilitates and enhances the formulation and implementation of programmes, incentives and projects on climate change (Council of Economic and Social Policy [CONPES], 2011). Four strategies were signalled: the National Adaptation Plan, the National Strategy for Reducing Emissions from Deforestation and Forest Degradation, the Strategy for Financial Protection against Disasters and the Colombian Low Carbon Development Strategy (CLCDS). The CLCDS is the most relevant in this case because it aims at identifying mitigation actions, including its abatement potentials and costs, to design sectorial mitigation plans and to implement them. Therefore, the CLCDS supports the formulation of NAMAs in different sectors, but until the date urban electric transportation is not part of the most prominent NAMAs (Colombian Low Carbon Development Strategy [CLCDS], 2013).

The use of electricity from renewable sources in the transportation sector could be an important measure to reduce GHG emissions and at the same time may contribute to one of the Colombian energy policy objectives which aims to diversify the energy consumption baskets. In that context, this paper aims to contribute with the characterization of that measure by two ways: the first one is to propose a methodological approach to assess the economic and environmental aspects of the use of electricity in the transportation sector; the second one is to perform a case study for the biggest city of the country in order to identify and characterize a possible NAMA (s) that might be implemented.

In particular, the proposed valuation methodology relies mainly on the model Model for Economic and Environmental Assessment of Electric Vehicles (MEEAVE) that was developed to assess the economic and environmental impacts of the introduction of electricity as an energy alternative to meet the transport needs of a region. This model allows to compare the costs faced by demand to meet its transport requirements with several fuel baskets in the different transportation modes (i.e. taxis, private passenger vehicles, public transport and cargo service). At the same time, the model allows to estimate the mitigation of GHG and particulate matter (PM), the changes in the electricity and fossil fuels consumption and the sales or fleet paths that must be followed in order to reach a proposed penetration goal. Results from the case study allow to identify the economic savings of electricity penetration in taxis and small urban freight services. On the other hand, private passenger vehicles and buses imply additional costs for fossil fuels substitution. The former represents the biggest opportunity for GHG mitigation among the different transportation sectors in Bogotá and the latest is the only service that could be centrally planned by major companies or by the city's administration.

This document has four sections hereafter. The first one briefly presents the Colombian energy and emissions situation, focusing on the transportation sector. The following two sections present the MEEAVE model and the case study for the city of Bogotá. Finally, the last sections are used to interpret and analyse the results and to present the final conclusions.

2. Energy and emissions context

This section contains a brief overview of the Colombian energy and GHG emissions context, with special emphasis on the transportation sector. Summarizing, Colombia has enough fossil fuel reserves and renewable resources (most of them hydro resources) to support its development process for the next years; energy is the second largest GHG emission sector and transportation is the main energy consumer, implying that it is mainly responsible for the emissions associated with the use of energy in the country.

In 2011, the country had proven oil reserves totaling 1.4 billion barrels, 60% of which were light oil (Energy and Mining Planning Unit [UPME], 2005). Proven reserves of natural gas in 2008 were 4.5 Tera cubic feet (Promigas, 2010). Regarding coal, the proven reserves in 2010 were 6814 million tonnes (British Petroleum, 2011). Colombian electricity generation is characterized for having an important hydro component that in 2011 produced 78% of the total national electricity (UPME, 2012). Colombia is the third largest regional consumer of hydropower after Brazil and Venezuela (British Petroleum, 2011), and has a theoretical hydro potential of 90,000 MW. The Colombian generation basket is complemented primarily with natural gas, to a lower extent with coal and some marginal participation of other type of plants (biomass and wind power plants). Moreover, the electricity sector experienced a reform in the early 1990s that enhanced the robustness and reliability of the supply of electricity.

Final energy demand in Colombia was 962.3 PJ in 2010 and the transportation sector accounted for 38% of that energy. Diesel is the main fuel used in road transportation followed by gasoline; by national regulations, both of them contain a percentage of biofuel (ethanol for gasoline and biodiesel for diesel). Compressed natural gas (CNG) was initially used only by public urban passenger transportation vehicles (taxi cabs and buses) and in recent years there has been an increasing number of vehicle conversions (automobiles and small trucks) from gasoline to CNG due mainly to high oil prices. As a result, CNG consumption has increased.

Other important changes that have taken place recently are the net increase in the automotive fleet and the incorporation of some bus rapid transit (BRT) systems in the main urban centres of the country. In 2010, there were approximately 3.6 million vehicles of which freight represents 6.7% and passenger transportation the remainder.

In local pollution, urban transport has been identified as the main cause of air pollution in urban centres of the country, this being the problem with the higher environmental and social costs associated, after water pollution and natural disasters (MAVDT, 2011). As a consequence, the government has made efforts to achieve the reorganization and optimization of the transport service. With respect to electric transportation, Decree 2439 of 2010 granted a tariff reduction for the import of clean technology vehicles, which applies to hybrid, electric and dedicated natural gas vehicles; and Decree 677 of 2011 ordered measures to encourage the use of electric vehicles in Bogotá and promote a pilot project for taxis during three years.

Estimations performed by Universidad de los Andes indicate that the scenario without constraints would lead to a fleet with 7.3 million vehicles and 9.4 million motorcycles in 2030. The forecast for population in that year is 56 million inhabitants (Echeverry Bocarejo, Acevedo, Lleras, Ospina, & Rodríguez, 2008). Currently, the rate of vehicles per 1000 people is 130, which is small compared to many other countries. On the other hand, the number of buses per person in the main cities of Colombia

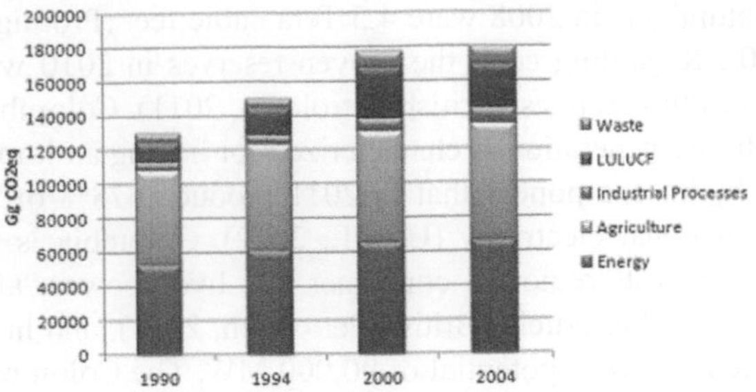

Figure 1. Total emissions in 1990, 1994, 2000 and 2004 (IDEAM, 2009).

(3.1 per 1000 people) is much higher than in Latin American cities such as Santiago, Curitiba or Quito.

According to Colombian Institute of Hydrology, Meteorology and Environmental Studies (Institute of Hydrology, Meteorology and Environmental Studies [IDEAM], 2009), the total GHG emissions in the Colombian transportation sector for 2005 were 20,000 Gg CO_2e. In 2004, energy use was responsible for 36.65% of the national emissions as can be seen from Figure 1. This sector is the second largest emitter of GHG in the country. Within the energy module, transportation contributed with 33% of the total emissions. The automotive subsector has the biggest share in the transport emissions, in 2004 emitting 89.5% of total emissions. Passenger transport is responsible for about a half of all emissions within the automotive subsector, of which about two-thirds are related to the urban passenger transportation. The above shows that urban passenger transportation (public and private) carries the main responsibility for the emissions of one-third of the total emissions of the energy module, which means 12% of the national emissions in 2004. In that order of ideas, in search of GHG abatement opportunities, the urban transportation sector should be considered as it is one of the main emitters within the Colombian context and it might also have some of the biggest abatement potential.

3. Methodological proposal. Model for economic and environmental assessment of electric vehicles – MEAVE

In order to perform the evaluation raised in this article, it is of paramount importance to review and take into account some of the models that have been implemented and are focused on the forecast and evaluation of electric vehicles. The models above can be classified into three broad categories, namely simulation, accounting and optimization. The simulation models are intended to analyse the behaviour of a system under certain initial conditions and parameters defining time variations. Accounting models focus on the quantitative estimation of parameters of environmental, economic or energy nature, affected by various inputs in a given time horizon; with these models, it is possible to synthesize the results by using statistics and also to identify relationships and dependencies. Optimization models compare various cases and determine the optimal value by considering some selection criteria.

By reviewing literature, various modelling options were found, whose features are quite extensive. Some of the models and software applications identified are: ADVISOR – ADvanced VehIcle SimulatOR – which implements a simulation and optimization model utilized for the analysis of performance, fuel economy and emission of conventional, electric, hybrid and fuel cell vehicles (Gao, Mi, & Emadi, 2007); Powertrain System Analysis Toolkit is simulation software that allows to structure the basic vehicle characteristics and select different configurations, predicting in an accurate way the fuel economy and performance (Gao et al., 2007); Long-range Alternatives Planning System is a software tool utilized to approximate energy consumption, production, resource extraction and emissions of local and regional air pollutants in all economic sectors (COMMEND) and it has been the primary tool (accounting model) used for the modelling of the penetration of electric vehicles in Thailand (Saisirirat Chollacoop, Tongroon, Laoonual, & Pongthanaisawan, 2013), by considering the changes in the behaviour of the load curve for electricity demand from vehicles; LEAP has been used in another case presented in Juan González (2012), where scenarios were developed to analyse the evolution of light-duty vehicle fleet in Colombia under different choices regarding powertrains, fuels and materials for vehicle manufacturing; a simulation model called VECTOR21, which compares scenarios, technologies and components, includes 3 sizes of vehicles, 10 types of technologies and 900 different users with defined preferences regarding vehicle attributes. The scenarios developed are intended to assess the impact of these vehicles on the market and it has been validated with historical information from the passenger fleet in Germany (Propfe, Kreyenberg, Wind, & Schmid, 2013); the Energy Economics Group of the University of Technology analyses the effects of policy, fuel prices and technological progress on the Austrian car fleet's energy consumptions and GHG emissions using an accounting model mathematically described in (Kloess & Müller, 2011); the University of Paris developed an accounting model to estimate environmental and economic trends that promote electricity to be used as alternative fuel in vehicles (Prud'homme & Koning, 2012); another accounting model was developed in the University of Oxford, called Transport and Carbon Simulation Model and it functions as an interactive game where the objective is focused on reducing emissions in the transport sector. This model considers the implementation of alternative fuels, policies, prices, differences in emissions on vehicles and different user roles ('free riders', 'techno-optimist', 'enviro-optimist' and the like). In (Hickman, Ashiru, & Banister, 2010) different packages or combinations that reduce and compare the levels of contamination are addressed and, finally, the International

Energy Agency – has a roadmap that includes a MARKAL model to identify the optimal conjunction of technologies and fuels to meet energy demand, taking into account constraints such as the availability of resources (Australian Energy Market Commission [AEMC], 2012). The model and the technique used in each analysis depend on the objectives of the study, the accuracy of the information and the flexibility of the interface used to enter data, simulate and analyze results.

In Colombia, the variety of models for assessing aspects of electric vehicles is not very extensive. A study case has been developed in the LOGIT model, a simulation model where the comparison of technology attributes (colour, price and performance) determines the probabilities of choice. Consequently, the impact on energy demand when penetrating electric and hybrid vehicles in the Colombian fleet is estimated (Zapata, 2009). The National University of Colombia used the MARKAL model to optimize technological participation with information based on demand and economic, energy, environmental and technological parameters (Universidad Nacional de Colombia – Sede Medellín, 2007).

MEEAVE emerges as an alternative model evaluation against electric vehicles, providing the user with a global perspective that includes the main technical, environmental, economic and energy features associated with the transportation sector. MEEAVE is composed of a friendly graphical interface and its concrete structure offers the possibility of prospective evaluations and comparisons between vehicles of different technologies, adjustable to any segment of the vehicle fleet or transportation system of any city.

3.1 *Model for Economic and Environmental Assessment of Electric Vehicles*

MEEAVE (acronym in Spanish for 'Model for Economic and Environmental Assessment of Electric Vehicles') was developed in this study in order to assess the penetration of electric vehicles in the road transportation sector. MEEAVE is an accounting model that was designed and built aiming to support decision-making processes tied to a specific set of modelling considerations agreed upon after discussions with decision-makers of some private electricity utilities and relevant public entities such as MADS and the Bogota's secretary office of transportation. Among the main modelling considerations assessed by MEEAVE are: electric vehicles technology is still under development, so prices are expected to decrease; it is desirable that the model allows to evaluate different types of goals and penetration trajectories of the new technology in a simple manner to facilitate comparison between measures; energy prices and electricity-emission factors might change during the evaluated period; sales are due to the requirements to replace scrapped old vehicles and increase the net fleet; batteries in electric vehicles represent an important share of full investment and it is expected that those vehicles may need to replace the batteries within their useful lifetime. Next, the structure of the model is summarized as well as the way in which some of the modelling considerations were addressed.

The model consists of five modules and allows an analysis in a time horizon of 30 years within a specific subsector (i.e. private vehicles, buses, taxis and light urban freight). The first module corresponds to the technical and economic characterization of the different transport technologies suitable for the relevant subsector. Vehicles are characterized by their main fuel; their investment and maintenance costs; average consumption and expected lifetime. As electric vehicle technology is still under development, it is expected that prices will decrease during the coming years; this module allows the incorporation of a changing technology price path and differentiating the prices of the vehicle and the required batteries.

In the second module, relevant energy sources are characterized by identifying their cost path and emissions associated with their use. Emissions associated with the use of electricity are nil, but there are some GHG emissions due to electricity production; those emissions depend on a specific electricity basket. Indeed, this module allows choosing whether or not GHG emissions associated with electricity production are accounted for; furthermore, the associated electricity emission factor might be modelled as a fix value or, if major changes in the electricity basket are expected, as a yearly changing value.

In the third module, transportation demand to be met is characterized, and a baseline of technologies used to satisfy that demand is established. Demand is defined by the yearly fleet size and by the average mileage. Demand projection is an input of the model and must be calculated exogenously. As electricity may be part of the baseline (e.g. when there are ongoing pilot programmes) this module allows the inclusion of a share of electric vehicles on it.

The fourth module allows the determination and characterization of electric fleet penetration scenarios to be compared against the baseline. This is one of the most prominent differences between MEEAVE and other energy-accounting models: as this model was designed specifically to evaluate electricity penetration in transport, it allows the evaluation of different types of goals and penetration trajectories in a simple manner, and simultaneously considering the replacement of vehicles and batteries when they reach their useful period or a given number of changing cycles. The goal that is being sought can be expressed as a percentage of annual sales of new vehicles, the participation in new vehicle sales in the final year, the kilometres powered by electricity in a given year, the total fleet circulating in a given year or directly dimensioning the size of the electric fleet in each year of the study period. The selected goal determines the trajectory of the fleet in the scenario. In Section 3.1.1, each of the possible evaluating measures is introduced in greater detail. In this module, values for discount rate and for social costs of particulate material (PM) are chosen by the user. Social costs of PM refer to an economic assessment of the impacts that PM has on public health and people's productivity; that value differs from town to town and is exogenously calculated. MEEAVE uses a social cost of PM introduced by the user to assess the potential savings on public health and productivity reached by avoiding the emissions of those sorts of pollutants.

Finally, in the fifth module, results are shown. Some of the information presented in this module include: differences in costs to satisfy the demand between the baseline and the penetration scenario; energy consumption, disaggregated by type of fuel (including electricity); the reached abatement of GHGs and local pollutants (PM), yearly fleet and sales trajectory for each type of fuel.

3.1.1 *Penetration goals*

3.1.1.1 *Kilometres travelled in the final year.* It allows establishing the minimum participation of electric technologies in the useful demand (measured as kilometres travelled) of the transport service in a specific year. The trajectory of electric fleet (VEB) in year t is determined as the number of initial vehicles if t equals the initial year of the measure or from the following formula for the other years.

$$\text{Electric fleet}_t = \text{Electric fleet}_{t-1} + \frac{\left(\left(\text{goal} * \dfrac{\text{Km travelled ICE fleet}_{\text{final year}}}{\text{Km travelled VEB}_{\text{final year}}}\right) - \text{Initial year vehicles}\right)}{\text{final year} - \text{initial year}}.$$

3.1.1.2 *Participation in annual sales.* It allows setting a goal of a minimum percentage of participation of electric vehicles in annual sales during the evaluation period.

$$\text{Electric fleet}_t = \text{Electric fleet sales}_t + \left(\text{Electric fleet}_{t-1}\left(1 - \frac{1}{\text{Useful life VEB}}\right)\right),$$

$$\text{Electric fleet sales}_t = \text{goal} * \text{total sales}_t.$$

3.1.1.3 *Participation in the final-year sales.* It allows setting a goal of participation of electric technology in vehicle sales of the final year. This goal is reached after successive increases in the participation of sales from an initial value (0% participation of sales, currently).

$$\text{Electric fleet}_t = \text{Electric fleet sales}_t + \left(\text{Electric fleet}_{t-1} * \left(1 - \frac{1}{\text{Useful life VEB}}\right)\right),$$

$$\text{Electric fleet sales}_t = \text{Participation in sales}_t * \text{total sales}_t,$$

$$\text{Participation in sales}_t = \frac{\text{goal}}{\text{final year} - \text{initial year}} + \text{Participation in sales}_{t-1}.$$

3.1.1.4 *Participation in the final-year fleet.* This goal allows the assessment of the cost and impact of a strategy seeking to have a certain number of electric vehicles circulating in a given year.

$$\text{Electric fleet}_t = \text{Initial year vehicles} * \text{EXP}\left[\ln\left(\frac{\text{goal} * \text{ICE fleet}_{\text{final year}}}{\text{final year} - \text{initial year}}\right) * (\text{current year} - \text{initial year})\right].$$

3.1.1.5 *Penetration trajectory.* It allows the establishment of an ad hoc trajectory of the size of the electric vehicle fleet in the scenario of each of the periods of the study horizon.

3.1.2 *Mathematical formulation*

The mathematical formulation of the model developed is introduced below. Set I, composed of the different technologies that can be used to meet the transport demand of the category that is being studied (diesel, gasoline, dedicated natural gas, converted natural gas and electricity), is defined. The costs of the ith technology vary over time so it is important to identify the initial year of operation of each type of vehicle. Elements of set J are the years in which a determined technology vehicle can start operation. L is the set of recorded pollutants and T is the set of years that are being studied. It further has

$$J \subseteq T$$

M	annual technology maintenance
FU_t	usage factor in year t
C_i	consumption technology i
$P_{i,t}$	fuel price i in year t
r	discount rate
$II_{i,j}$	initial investment technology i in model j
VU_i	useful life technology i
$S_{i,t}$	share of technology i in total annual sales of year t
$D_{i,t}$	demand of technology i in year t
$FE_{i,l,t}$	factor of emission of pollutant l in technology i in year t
m	cost for society of PM in the atmosphere

Operation and maintenance costs of technology i in year t are given by

$$COM_{i,t} = (FU_t \cdot C_i \cdot P_{i,t}) + M_i.$$

The total cost of technology i entered in period j is calculated as

$$C_{t,j} = \sum_{k=0}^{VU_i - 1} \frac{COM_{i,j+k}}{(1+r)^k} + II_{i,j},$$

with which the annual value of technology i, entered in period j, is obtained

$$\mathrm{VA}_{i,j} = \frac{\mathrm{CT}_{i,j} \cdot r}{1-(1+r)^{-\mathrm{VU}_i}}.$$

The number of vehicles of technology i entered in period j and that are still in operation in year t is

$$f_{i,j,t} = \begin{cases} 0 & \text{si} \quad j+\mathrm{VU}_i < t < j \\ v_{i,t} \cdot \left[1 - \frac{t-j}{\mathrm{VU}_i}\right] & \text{si} \quad j+\mathrm{VU}_i \geq t \geq j \end{cases}.$$

The entry of vehicles of each i type is given as a result of sales over a period of time. Total sales in period t are calculated as

$$\mathrm{VT}_t = \sum_I \left(D_{i,t} - D_{i,t-1} + \frac{D_{i,t-1}}{\mathrm{VU}_i}\right),$$

with which it is possible to quantify sales of technology i in period t as

$$v_{i,t} = VT_t \cdot s_{i,t}.$$

Thus, the annual costs of meeting the demand for transport of the category evaluated are

$$\mathrm{CAT}_t = \sum_I \sum_J f_{i,j,t} \cdot \mathrm{VA}_{i,j}.$$

And the total cost of meeting demand in the time period under study is

$$\mathrm{CT} = \sum \frac{\mathrm{CAT}_t}{(1+r)^{t-t_0}}.$$

Fuel consumption is necessary for estimating the emission of pollutants and is equivalent to

$$CC_{i,t} = \mathrm{FU}_t \cdot C_i \cdot \sum_J f_{i,j,t}.$$

Emissions of pollutant l are estimated as

$$E_l = \sum_I \sum_T \mathrm{CC}_{i,t} \cdot \mathrm{FE}_{i,l,t},$$

with $l=2$ we have the cost of the PM as

$$\mathrm{CMP} = \sum_I \sum_T \frac{\mathrm{CC}_{i,t} \cdot \mathrm{FE}_{i,2,t} \cdot m}{(1+r)^{t-t_0}},$$

with which finally the total system cost is obtained deducting the benefits obtained from savings of PM:

$$\mathrm{CTC} = \mathrm{CT} - \mathrm{CMP}.$$

4. A case study: electric transportation for the city of Bogotá

For the specific case of Bogotá, a portfolio of electricity penetration in the transportation sector was designed and evaluated. For this purpose, urban transportation demand was divided into several categories: private transport, taxis, light-load fleet and Transmilenio mass transport system. For such categories, a specific goal was defined. Those goals as well as the main assumptions for the modelling process were agreed upon with Bogotá's electric utility (Codensa) with some guidelines from the Energy and Mining Planning Unit (UPME), the Bogotá's transportation secretary office and the MADS. The following subsections will elaborate further on the exercise performed: characterizing the technologies and the energy carriers considered, describing the demand for each sector, delineating the evaluated measures and presenting the obtained results.

4.1 *Electric-transportation technologies and energy carriers*

In order to evaluate the designed goals, we chose a standard vehicle that represents each category considered. Thus, each category is characterized by a single type of vehicle for each fossil fuel and by a single electric vehicle selected according to its characteristics in the currently available commercial cars. It is important to note that the information presented does not intend to advertise any brand of vehicles but corresponds to a selection made by authors based on a set of predefined criteria intended to find similar vehicles, in terms of capacity and body size, for each use category.

For private vehicles, two types of technologies were evaluated, private cars and station wagons. The car chosen as the standard for passenger cars is the Renault Fluence. The evaluation of the introduction of this type of vehicles in the fleet will take into account the participation of both, gasoline vehicles and converted from gasoline to natural gas vehicles. With respect to private station wagons, the vehicle selected is the Renault Kangoo. The technologies considered within the category mentioned above apply to vehicles using gasoline, natural gas (converted from gasoline) and diesel. Equipment considered in the evaluation of utility vehicles is the same as the one used to assess private station wagons; although as a result of their increased activity, annual maintenance costs for utility vehicles (cargo fleet) is 2.2 times the equivalent for the private station wagons. The technology selected for modelling gasoline- and natural gas-powered taxis circulating in the city is the Accent Vision model of Hyundai, and the equivalent electric vehicle considered is the E6 of BYD. Vehicles of the Transmilenio mass transport system were classified as articulated and standard. In both cases, diesel vehicles and electric vehicles were considered. For electric articulated buses we modelled catenary buses instead of battery buses; infrastructure costs were not taken into consideration for the analysis.

Goals shown in next sections were assessed in three fuel-price scenarios. Those scenarios were built based on the considerations presented by the US Energy Information Administration (2011). It was assumed that the relative price of fossil energy sources remained constant throughout the study horizon. On the other hand, prices for electricity were assumed constant, considering that, during the following years, there are no expectations of major changes in the composition of the electricity basket, which is currently based on hydro and local fossil resources. The assumption that the electricity basket will remain constant in its composition is based on the fact that unexploited hydro resources in the country are enough to support the growing requirements. The above also implies that the emission factor associated with the electricity production will remain rather constant in comparison to the current value (UPME, 2010a, 2010b). The values for the emission factor for fossil fuels were taken from the database used in IDEAM (2009).

4.2 *Evaluated portfolio*

Transportation demand of the base year was estimated based on the number of trips generated in the city and taking into account the different modes of transport that supply that demand. In 2008, an average of 12.2 million daily trips were made in Bogotá (CCB-Uniandes, 2009). Twenty-two per cent of the trips were made by private transportation (light vehicles, all-terrain vehicles, station wagons and motorcycles), 23% by non-motorized means of transport (bicycle and walking) and 55% by public transport (taxis, Transmilenio mass transport system and conventional collective transport system – buses and minibuses).

The total travel demand was projected using the function of mobility ratio (number of trips/per capita) developed in a previous study for Colombia (Echeverry et al., 2008). The growth of the private vehicle fleet was modelled with the motorization rate (number of vehicles/1000 per capita) projected to 2040. In 2008, there were 129 vehicles in the city per 1000 inhabitants; for 2040, a rate of 490 vehicles per 1000 inhabitants was projected. According to a study conducted for 46 countries in the world (Dargay, Gately, & Sommer, 2007), the motorization rate in the USA, Canada, Belgium, Germany, France and Britain in 2002 was already higher than the value estimated for Bogotá in 2040. Moreover, the value obtained for Bogotá in 2040 is similar to what was shown in such study for Israel, México and Argentina in 2030. The average annual growth rate (of the motorization rate) obtained for Bogotá between 2008 and 2040 is of the same order of the growth rate of Brazil, Ecuador and the Dominican Republic for the period 2002–2030, according to the results of the aforementioned study. With the assumptions used, the number of private cars will grow at an average annual rate of 5.8%, which means that in 2040 there will be about five million private vehicles.

In the demand projection of the reference scenario for public transport was considered the demand of the Transmilenio mass transport system with future works (six new BRT corridors). Moreover, the implementation of the Integrated Public Transport System (SITP, for its acronym in Spanish), its demand goals, scrappage of buses and integration with new modes of transport in the city, such as the metro and the light rail were also taken into account (Colombian Comptroller's Office, 2010; Econometría, 2007).

Finally, the projection of taxis and non-motorized modes of transport was estimated assuming that they gradually reduce their participation in city trips by 15%, following the rise of private transport and public transport provision. This is the projected reduction for these two modes for Bogotá in Echeverry et al. (2008). Values of annual mileage of the vehicles and fuel participation were obtained from a previous study developed for Bogotá (Universidad de los Andes, 2010a, 2010b).

In that context, a set of goals for each of the categories considered was formulated according to the market expectations perceived by experts of Endesa in Colombia. For the case of public buses, and considering that changes in the fleet (replacements or increase of the fleet) occur only in discrete periods, a trajectory that coincides with the expected expansion of the system was proposed, assuming that these expansions are made only with electric vehicles. For the other categories of vehicles (i.e. particular vehicles, freight and taxis), we set a goal of participation of electric vehicles in the final-year fleet, starting penetration in 2020. Values for that goal were defined to be 30% in taxis and light-load fleets and 15% in private cars. Next, results from assessing those goals using MEEAVE are presented.

4.3 *Quantitative results and qualitative analysis*

This section presents the main quantitative results of the case study. As one of the objectives of this study is to identify and characterize a possible NAMA that might be implemented, quantitative results are complemented with a qualitative analysis regarding actions required to allow reaching the identified abatement potentials as well as some guidelines aimed for a measurement, reporting and verification (MRV) system in the Colombian context.

Table 1 summarizes the main results obtained by modelling the proposed penetration portfolio. Those results will be further analysed in this section. Presented results are those obtained considering the intermediate scenario of fuel prices. Values for CO_2 abatement were calculated considering the increase in emissions due to increase in electricity production.

Table 1. Results from the evaluation of an electricity vehicle penetration portfolio in an intermediate oil price scenario.

	Taxis	Light freight	Particular vehicles	Particular passenger trucks	Articulated buses	Regular buses
Total substitution costs (thousands US$)	−6,514.3	−93,511.1	13,764.9	5,928.3	10,4652.7	30,258.1
CO_2 Abatement (Gg CO_2)	697.0	1,939.4	5,292.2	1,936.6	2,442.5	342.2
PM emission reduction (kg PM)	69,939.3	52.241.2	95,891.5	52,167.8	4,526.8	808.1
Savings by avoiding PM (thousands US$)	2,632.4	2,212.7	3,534.8	1,874.5	390.4	70.4
Total cost including savings by avoiding PM (thousands US$)	−9,146.7	−95,723.8	10,230.1	4,053.8	10,075.2	30,187.7
Cost of avoided CO_2 (US$/t$CO_2$)	−9.3	−48.2	2.6	3.1	4.3	88.4
Cost of avoided CO_2 including savings by avoiding PM (US$/t$CO_2$)	−13.1	−49.4	1.9	2.1	4.1	88.2

Figure 2 shows, as example, the conformation of the fleet of utility vehicles to a penetration goal of 30% in 2040; other categories present similar behaviours. That penetration contrasts with the behaviour of GHG emissions that are shown in Figure 3. As can be seen, before 2030 emissions of GHG reduce their growing rate but after that emissions reverse their growing rate. This is a very interesting result that reveals that in Colombia, due to the low emission factor in electricity production, a 30% of participation of electricity in the fleet decreases not only the growing rate of GHG emissions but also their absolute value in most of the categories. For example, freight fleet emissions in 2040 are less than emission levels reached in the same scenario (with electricity penetration) by 2032, which means that GHG emission growing rates during the last years were negative.

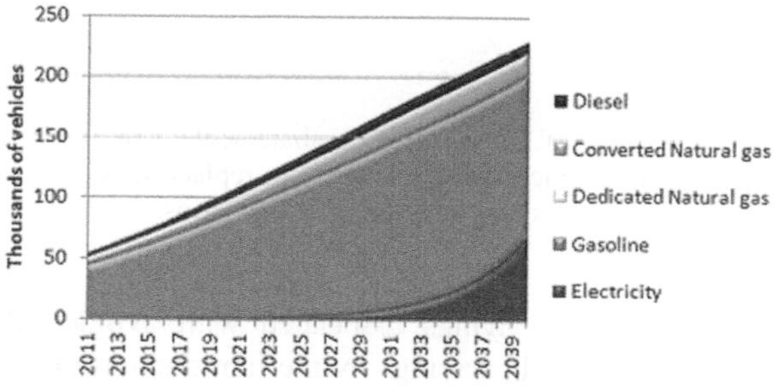

Figure 2. Composition of the light-load vehicle fleet by energy source.

Electricity consumption due to electrical transport penetration is presented in Figure 4. The maximum electricity consumption of the modelled sectors is reached in 2040: 2108.3 GWh-year. Considering that total demand for electricity in Colombia during 2010 was close to 52,000 GWh-year, it can be seen that the additional demand corresponding to electrical transport in Bogotá would be a small fraction of the national annual demand. In fact, consumption of modelled vehicles in 2040 represents 4% of the Colombian electricity consumption in 2010. In other words, the Colombian power sector will be required to increase its generation by a total of 4% in 30 years in order to fulfill the transportation demand in Bogota; that requirement could be satisfied easily by Colombian power sector without modifying its electricity basket. On the other hand, it still has to be considered how the additional demand is distributed throughout the day to assess the impacts that apparently seem insignificant in the transmission and distribution networks.

As mentioned above, substitution of natural gas and oil derivatives for electricity implies reduction of GHGs in Colombia, even considering the additional emissions caused by the increase in the country's electricity generation. Figure 5 shows the reductions of GHGs achieved by using the amounts of electricity presented in Figure 3

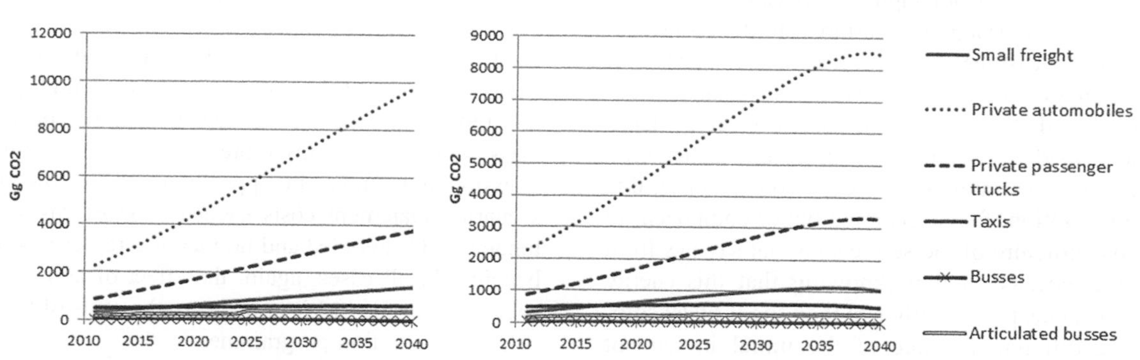

Figure 3. Emissions of carbon dioxide in the penetration scenario of electric vehicles in Bogotá. Without electricity penetration (left) and with electricity penetration (right).

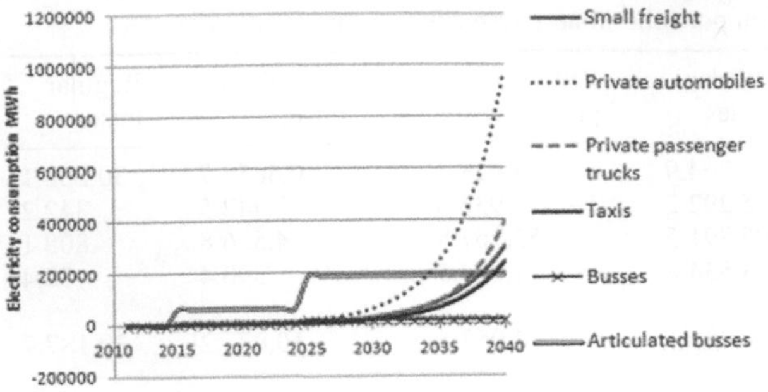

Figure 4. Yearly electricity consumption of the different technologies of electric transport in Bogotá in an electricity penetration scenario.

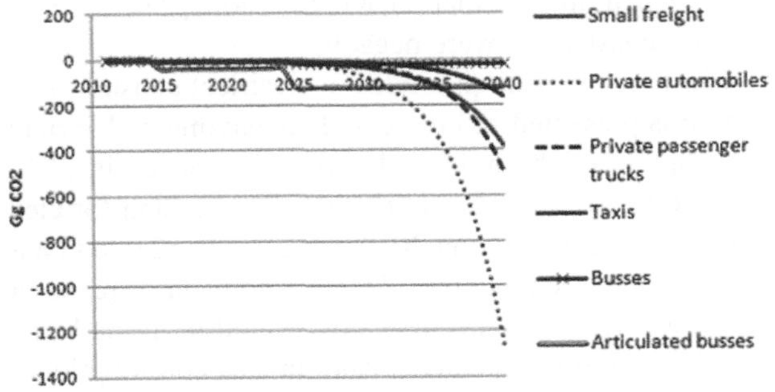

Figure 5. Reduction of carbon dioxide in an aggressive electricity penetration scenario in transportation in Bogota.

to replace natural gas and oil derivatives in transportation in Bogotá. By 2040, those reductions would represent net savings of 15.3% compared to the baseline (without electricity penetration). In all cases, the reduction of PM is at least equal in proportion to the reduction of GHGs, with corresponding savings in terms of health for society in general. It is interesting to note that, despite the fact that private transport has a lower level of annual activity, its consumption and reduction potentials are significantly higher than those of other types of transport. This can be explained by the differences in the sizes of the fleets in each type of service.

A summary of the economic analysis of the different measures can be seen from Figure 6. It can be seen that there are three main groups in relation with the cost effectiveness of the measures: non-regret measures (taxis and freight), 'affordable' measures (private vehicles and articulated buses) and 'expensive' measures (regular BEV buses). Light-load fleets and taxis show that penetration of electricity implies savings. In these two categories, due to its high utilization, fuel costs are a major component in the total cost structure of the service, so that savings from the use of electricity as fuel (knowing that this energy source is much cheaper and efficient than fossil fuels) compensates the differential value of the initial investment when buying an electric vehicle instead of a cheaper conventional car.

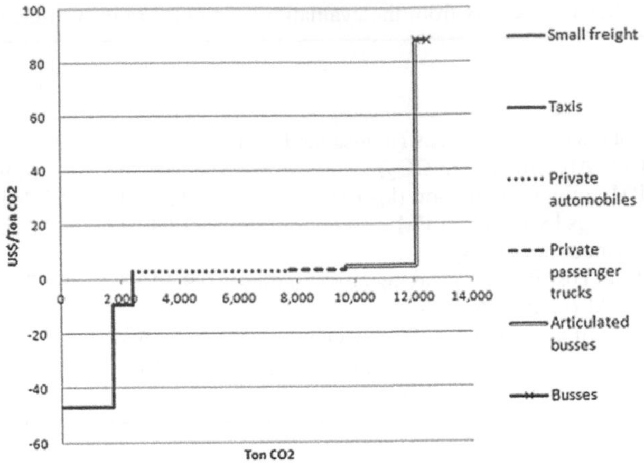

Figure 6. Costs of carbon dioxide abatement in the transport sector in Bogotá when using electricity to replace fossil fuels in a horizon of 30 years in an average oil price scenario.

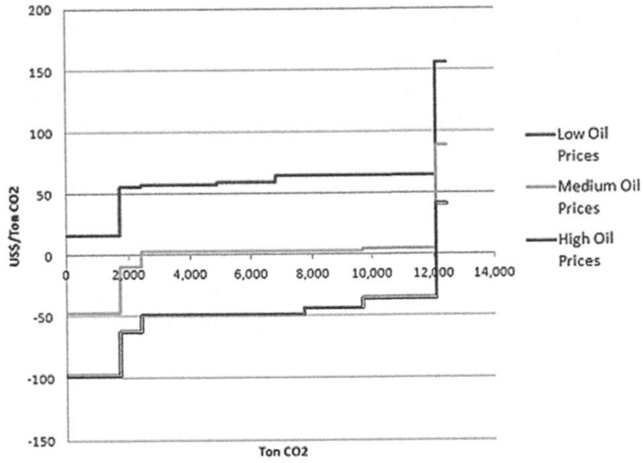

Figure 7. Costs of carbon dioxide abatement in the transport sector in Bogotá when using electricity to replace fossil fuels in a horizon of 30 years and three oil price scenarios.

If considered, savings associated with avoided PM increases by 40% the savings by the introduction of taxis and decreases by 26% and 32% the additional costs of use of electricity in private cars and station wagons (as shown in Table 1). On the other hand, the same concept modifies the total costs of benefits in the other sectors by less than 5%.

These results are heavily dependent on oil prices, as shown in Figure 7. In a scenario of future high prices for oil, almost all measures (except the regular buses of the mass transport system) represent monetary savings and emission reductions. In opposition, in the low oil price scenario, abatement costs exceed in most cases \$50 per tonne of CO_2 avoided and no measure represents financial benefits. In all cases, again, the effect of considering the social cost of the PM is to decrease the costs of the electric vehicle penetration programme.

An electricity penetration portfolio, as the one proposed here, may be considered an appropriate mitigation action.

Electric vehicles are a plausible development strategy for the country considering that diversification of the energy basket has benefits in terms of security of supply and that PM reductions have positive impacts in the population health (and its productivity); furthermore, the increase in electricity demand as a result of using this energy source in transport does not seem to be important from the point of view of requirements of installed generation capacity. As Colombian electricity is generated (and it is expected that it will continue being generated) mainly from renewable sources, its use replacing oil derivatives and natural gas in transportation allows GHG abatements calculated in this study.

However, there are some barriers that must be overcome in order to allow measures like those proposed in this paper to become reality. The first of them is related to infrastructure requirements such as charging stations. Penetration of electric vehicles can be delayed due to the lack of public charging stations or due to the costs of the home charging stations. The second barrier is related to economic aspects. Results from the model identify the economic benefits of electricity penetration in taxis and small urban freight services; savings on fuel are enough to offset the initial investment costs. However, while additional investment costs must be faced at the time of buying the vehicle, savings are reached throughout the vehicle's useful life. This barrier could be overcome by enabling lines of credit for the down payment. On the other hand, private passenger vehicles and buses imply additional costs. Private passenger vehicles represent the biggest GHG mitigation potential. Their mitigation potential can be used to obtain financial support in order to reduce the gap between the traditional internal combustion vehicles and the electric ones. Finally, electricity use in the public massive transportation system is the most expensive alternative and does not have a significant GHG mitigation potential. In contrast, that is the only transportation sector that could be centrally planned by major companies (bus owners or Transmilenio) or by the district administration, so its implementation might be defined by political initiative.

Thus, a NAMA to incentivize electricity penetration in Bogota's transport sector may be structured to find financial support in order to overcome the barriers mentioned above (or others). In any case, a NAMA requires the implementation of a MRV system. Next, some guidelines about those requirements will be shown in order to contribute with a starting point for a possible NAMA that might be implemented to support the electrification of Bogota's transportation looking for the GHG abatements identified in this study.

4.3.1 *Guidelines for an MRV system for a possible electric vehicles NAMA for Bogota*

MRV systems are considered as fundamental in NAMA's context (Fransen, 2009; McMahon, & Moncel, 2009). Despite the MRV systems not yet being defined by the UNFCCC, some approaches to MRV systems have been proposed by different authors (Clean Air Asia, 2011; Center for Clean Air Policy, 2011, 2012; Fransen, 2009; Fukuda, 2009; GIZ, 2013; Höhne, Jung, Ward, & Ellermann, 2007; McMahon, & Moncel, 2009; Tilburg, Röser, Hänsel, Cameron, & Escalante, 2012).

An MRV structure could be based on the existing verification system that the Colombian government uses in order to verify the accomplishment of the goals proposed in the NDP. Within this scheme is the SIGOB, which is an information system implemented by the National Planning Department and the Presidency. It consists of verifiable numerical targets and indicators. The evaluation of the NDP is reported periodically to ministries, administrative departments, sector agencies, council of ministries, the National Planning Council and the National Congress (NPD, 2010a, 2010b). For specific indicators, we found useful the National Traffic Unique Registry (RUNT) from the Transport Ministry and some of the reporting and verification systems that are currently in use in institutions, for example, the ones used for the accomplishment of ISO standards (e.g. 9000 or 14,001).

Considering the mitigation measures previously analysed, some indicators were selected. Tables 2–4 introduce

Table 2. Implementation and monitoring indicators.

Type of indicator	Indicator	Units	Responsible entity
Implementation	Number of electric vehicles in public transport systems	Number of vehicles/year	Transport Ministry
Implementation	Number of electric vehicles in taxi fleets	Number of Vehicles/year	Transport Ministry
Implementation	Total electric vehicles registered each year	Number of vehicles/year	Transport Ministry
Monitoring	Energy intensity of the urban public transport sector (including BRTs)	MBTU/(passenger-Km) or GWh/(passenger-Km	National Energy Planning Unit, BRT companies and Transport Ministry. The indicator could be built by the Environment Ministry
Monitoring	Passenger Kilometres Index in electric BRT systems	Passenger/kilometre	BRT companies and the Transport Ministry

Source: Prepared by the authors.

Table 3. Indicators from National Development Plan.

Type of indicator	Indicator	Units	Responsible entity
NDP Programme: Environmental management and urban sector	Vehicles entering the fleet with cleaner technologies (natural gas, hybrids and electric vehicles)	Number of vehicles/year	Environment Ministry
NDP strategy: housing and friendly cities	Number of cities with urban mobility solutions in operation	Number of cities/ year	Transport Ministry
NDP strategy: equal opportunities for social prosperity	Adjusted infant mortality rate, per 1000 born	Percentage	Social Protection Ministry

Source: NPD (2011a, 2011b).

Table 4. Additional monitoring indicator.

Type of indicator	Indicator	Units	Responsible entity
Sustainable development	Concentration of PM in urban centres where NAMA is implemented	$\mu g\ PM_{10}/m^3$	Environment Ministry

Source: Prepared by the authors.

the indicators with a brief description. It includes possible actors in charge of measuring and reporting them.

Even though information in the country is in general limited (National Administrative Department of Statistics, 2006), we considered that some of the existing systems are useful in reporting and monitoring indicators for an electric transport NAMA, at least as a starting point. Some indicators could be used just as they are designed, while some others can be estimated using data already being monitored. Moreover, using the existing systems in NAMA's context can also contribute to overcome the actual limitations on information in the country, and this in turn may facilitate designing and implementing other low-carbon projects.

5. Conclusions

The model (MEEAVE) developed in this paper proved to be a flexible tool to evaluate different penetration targets of electric vehicles and useful to design the pilot programme for the city of Bogota. MEEAVE allows the user to consider different technologies as well as different strategies to reach penetration goals. Its modular configuration drives the user through five different modules starting with a detailed techno-economic representation of vehicles technologies as well as prices and environmental emission paths for energy carriers considered in the model. The model allows estimating transportation demand for different road modes. Finally, different electric fleet penetration scenarios can be evaluated: as a percentage of new vehicles sales or total circulating fleet, as a target of total kilometres travelled with electricity power or replacement of existing vehicles and batteries, or directly dimensioning the size of the electric fleet in each year of the study period. The selected goal determines the trajectory of the fleet in the scenario. Results are displayed in a simple manner: differences in energy consumption, costs, GHG and PM emissions to satisfy the demand between baseline and the penetration scenario as well as yearly fleet and sales trajectory for each type of fuel.

Results from this study were used to conceive a pilot programme for the city of Bogota. A portfolio of electricity penetration in the transportation sector was designed and evaluated. Results drive ENDESA and the MADS to start structuring a NAMA under the Mitigation Actions Plans and Scenarios Initiative. NAMA is an interesting alternative, taking into account the country's context, the goals of the NDP and the Energy Plans. Electric vehicles are a plausible development strategy for the country. Colombian electricity is relatively clean; therefore, substitution of fossil fuels by electricity represents net savings in emissions of GHG in the country. Even though, the increase in electricity demand as a result of using this energy source in transport does not seem to be important from the point of view of requirements of installed generation capacity, the diversification of the energy basket has benefits in terms of security of supply.

As far as the NAMA proposal is concerned, we can say that the exercise of structuring a NAMA from scratch is not an easy task. The identification of technology, the construction of base lines and the definition of scenarios were quite time consuming and required a wider participation of the utilities and national and local energy and environmental authorities. In this document, first steps required for the structuration of a NAMA were taken (i.e. identifying abatement potential, development benefits, economic costs and possible barriers for implementation) as well as a preliminary proposal for an MRV system.

One of the main limitations of the model is that, being of annual resolution, it does not allow the identification of daily charging patterns, which is critical from the point of view of the distribution network. It is suggested as future work to complement the model or use an additional tool in order to overcome that limitation. Another task to be assessed in the future refers to the identification of regulatory requirements and barriers that a portfolio like the one proposed in this document must face. Finally, the case study must be analysed vis-à-vis sensitivity to other parameters different from oil prices. In fact, the evaluated portfolio was only one of several portfolios considered and sensitivities were not further developed here because it was out of the scope of this paper, which aims mainly to present a methodological proposal and the evaluation of the most likely portfolio in the most probable scenario (according to criteria of Bogota's electric utility and MADS experts).

References

Clean Air Asia. (2011). *Transport data in Asia study* (Unpublished).

Australian Energy Market Commission, AEMC. (2012, August). *Energy market arrangements for electric and natural gas vehicles – Summary of AECOM's final advice. Obtenido de.* Retrieved from http://www.aemc.gov.au/Media/docs/Information-Sheet--Summary-of-AECOMs-final-report-9469211a-dda9-43d3-8dc6-48077055e3d4-0.pdf

Baumert, K., Herzog, T., & Pershing, J. (2005). *Navigating the numbers. Greenhouse gas data and international climate policy.* Washington, DC: World Resources Institute.

British Petroleum. (2011). *Statistical review of world energy 2010.* London. Retrieved from http://www.bp.com/assets/bp_internet/globalbp/globalbp_uk_english/reports_and_publications/statistical_energy_review_2011/STAGING/local_assets/pdf/statistical_review_of_world_energy_full_report_2011.pdf

Cadena, A., Salazar, M., Delgado, R., Espinosa, M., Rosales, R., & Rojas, A. (2011). *Country study of Colombian Mitigation actions.* Research report. Mitigation Action Plans and Scenarios (MAPS programme). Retrieved from http://www.mapsprogramme.org/wp-content/uploads/Mitigation-Action-Case-Study_Colombia.pdf

CCB-Uniandes. (2009). *Chamber of Commerce of Bogotá-Universidad de los Andes, Observatorio de Movilidad.* 2009. Bogotá, Colombia. Retrieved from http://camara.ccb.org.co/contenido/contenido.aspx?catID=127&conID=4905

Center for Clean Air Policy. (2011). *MRV for Namas: Tracking Progress while Promoting Sustainable Development.* Retrieved from http://ccap.org/assets/MRV-for-NAMAs-Tracking-Progress-while-Promoting-Sustainable-Development_CCAP-November-2011.pdf

Center for Clean Air Policy. (2012). *MRV of NAMAs: Guidance for Selecting Sustainable Development Indicators. Discussion draft.* Retrieved from http://ccap.org/assets/MRV-of-NAMAs-Guidance-for-Selecting-Sustainable-Development-Indicators_CCAP-Oct-2012.pdf

Colombian Low-Carbon Development Strategy. (2013, April). *Avances Estrategia Colombiana de Desarrollo Bajo en Carbono – Abril 2013.* Recuperado el June de 2013, de Colombian Ministry of Environment and Sustainable Development. Retrieved from http://www.minambiente.gov.co/documentos/DocumentosGestion/cambio_climatico/estrategia_bajo_carbono/100713_avances_ecdbc_abril.pdf

Colombian Comptroller's Office, Bogotá. (2010). *Viabilidad financiera del proyecto metro, evaluar la articulación metro-SITP. Plan anual de estudios 2010.* Bogotá, Colombia: Department of District Treasure and Finances.

Community for Energy, Environment and Development, COMMEND (s.f.). *LEAP, Long-range Energy Alternatives Planning System. Recuperado el 2013, de.* Retrieved from http://www.energycommunity.org/

Council of Economic and Social Policy. (2011). *Documento CONPES 3700. Estrategia institucional para la articulación de políticas y acciones en materia de cambio climático en Colombia.* Bogotá: National Planning Department.

Dargay, J., Gately, D., & Sommer, M. (2007). Vehicle ownership and income growth, worldwide: 1960–2030. *Energy Journal, 28*(4), 163–190.

DOE, & EIA. (2011). *International energy outlook.* Washington, DC: US Department of Energy, Energy Information Agency.

Echeverry, J., Bocarejo, J., Acevedo, J., Lleras, G., Ospina, G., & Rodríguez, A. (2008). *El transporte como soporte al desarrollo del país. Una visión al 2040.* Universidad de los Andes, Bogotá.

Econometría, S.A. (2007). *Evaluación de las posibilidades de utilización de medios de transporte energizados con electricidad. Informe final.* Bogotá, Colombia. Retrieved from http://www.si3ea.gov.co/Portals/0/URE/finelect.pdf

Energy and Mining Planning Unit. (2005). *La cadena del petróleo en Colombia.* República de Colombia, Ministerio de Minas y Energía, Unidad de Planeación Minero Energética: Bogotá, Colombia. Retrieved from http://www.upme.gov.co/Docs/Cadena_Petroleo_2004.pdf

Energy and Mining Planning Unit. (2010a). *Plan de Expansión de referencia – Generación y transmisión 2010–2024.* Bogotá, Colombia: República de Colombia, Ministerio de Minas y Energía, Unidad de Planeación Minero Energética. Retrieved from http://www.upme.gov.co/Docs/Plan_Expansion/2010/Plan_Expansion_2010-2024_Definitivo.pdf

Energy and Mining Planning Unit. (2010b). *Proyección de demanda de energía en Colombia.* Bogotá, Colombia: República de Colombia, Ministerio de Minas y Energía, Unidad de Planeación Minero Energética. Retrieved from http://www.siel.gov.co/siel/documentos/documentacion/Demanda/proyeccion_demanda_ee_Abr_2013.pdf

Energy and Mining Planning Unit. (2012). *Reporte mensual de evolución de variables de generación. Mayo 2012.* Bogotá, Colombia: República de Colombia, Ministerio de Minas y Energía, Unidad de Planeación Minero Energética.

Fransen, T. (2009). *Enhancing today's MRV framework to meet tomorrow's needs: The role of national communications and inventories.* WRI Working Paper, *Washington.* Retrieved from http://www.undpcc.org/docs/UNFCCC%20negotiations/Countrie%20positions%20and%20analysis/

Fukuda, K. (2009). *A 'measurable, reportable and verifiable (MRV)' framework for developing countries.* IGES Briefing Notes on the Post-2012 Climate Regime. Issue 10. Hayama, Kanagawa, Japan: Institute for Global Environmental Strategies. Retrieved from http://pub.iges.or.jp/modules/envirolib/upload/2878/attach/mrv_3rd.pdf

Gao, D.W., Mi, C., & Emadi, A. (2007). Modeling and simulation of electric and hybrid vehicles. *Proceedings of the IEEE, 95*(4), 729–745. doi:10.1109/JPROC.2006.890127

GIZ. (2013). *Measurement, reporting and verification (MRV) of NAMAs.* Presentation for the "Regional workshop on promoting international collaboration to facilitate preparation, submission and implementation of NAMAs", Maseru, Lesotho. Retrieved from http://unfccc.int/files/cooperation_support/nama/application/pdf/mrv_of_namas_-_giz.pdf

Hickman, R., Ashiru, O., & Banister, D. (2010). Transport and climate change: Simulating the options for carbon reduction in London. *Transport Policy, 17*(2), 110–125.

Höhne, N., Jung, M., Ward, M., & Ellermann, C. (2007). *Sectoral proposal template: Transport. Technical report.* Global Climate Change Consultancy (GtripleC) and Ecofys.

Institute of Hydrology, Meteorology and Environmental Studies.| (2009). *Inventario Nacional de Fuentes y Sumideros de Gases de Efecto Invernadero 2000–2004.* Bogota. ISBN:

978-958-96863-7-9. Retrieved from https://documentacion.ideam.gov.co/openbiblio/Bvirtual/021471/021471.htm

Juan González, T.F. (2012). *Low carbon road passenger transportation system design using combined energy and materials flows in Colombia*. Sendai: Tohoku University.

Kloess, M., & Müller, A. (2011). Simulating the impact of policy, energy prices and technological progress on the passenger car fleet in Austria – a model based analysis 2010–2050. *Energy Policy, 39*(9), 5045–5062.

MAVDT. (2011). *Colombia CDM Portfolio 2011. Technical report*. Bogotá, Colombia: República de Colombia, Ministerio de Ambiente, Vivienda y Desarrollo Territorial.

McMahon, H., & Moncel, R. (2009). *Keeping track: National positions and design elements of an MRV framework*. WRI Working Paper. Washington. Retrieved from http://www.undpcc.org/docs/UNFCCC%20negotiations/Countrie%20positions%20and%20analysis/WRI_KeepingTrack_June2009.pdf

National Administrative Department of Statistics. (2006). *Colombian national strategic plan for statistics. Comunidad Andina*. Retrieved from http://estadisticas.comunidadandina.org/eportal/contenidos/imagenes/file/pendes/docs/Cartilla_PENDES_colombia.pdf

NPD. (2010a). *Prosperidad para todos. Más empleo, menos pobreza y más seguridad. Bases del Plan Nacional de Desarrollo 2010–2014*. Bogotá, Colombia. República de Colombia, Departamento Nacional de Planeación.

NPD. (2010b). *Evolución de SINERGIA y evaluaciones en administración del Estado*. Bogotá: Dirección de Evaluación de Políticas Públicas, Departamento Nacional de Planeación. Retrieved from https://sinergia.dnp.gov.co/sinergia/Archivos/Cartilla_UNO_Sinergia_Admon_Estado.pdf

NPD. (2011). *CONPES 3700 de 2011. Cambio climático*. Bogotá, Colombia.

NPD. (2011a). *Un cuatrienio de oportunidades, crecimiento e innovación. [Powerpoint slides]*. Retrieved from https://www.dnp.gov.co/LinkClick.aspx?fileticket=8g7YTLGNNOo%3d&tabid=82

NPD. (2011b). *Plan Nacional de Desarollo 2010–2014 Prosperidad para todos*. Colombia. Bogotá: Departamento de Planeación Nacional de Colombia.

Promigas. (2010). *Informe del sector gas natural 2010. Un balance de la década*. Retrieved from http://www.promigas.com/wps/wcm/connect/84a9ad8048997f5aabf7ebcfb9c50a54/Informe+del+sector-2011.pdf?MOD=AJPERES&CACHEID=84a9ad8048997f5aabf7ebcfb9c50a54

Propfe, B., Kreyenberg, D., Wind, J., & Schmid, S. (2013). Market penetration analysis of electric vehicles in the German passenger car market towards 2030. *International Journal of Hydrogen Energy, 38*(13), 5201–5208. doi:10.1016/j.ijhydene.2013.02.049

Prud'homme, R., & Koning, M. (2012). Electric vehicles: A tentative economic and environmental evaluation. *Transport Policy, 23*, 60–69.

Saisirirat, P., Chollacoop, N., Tongroon, M., Laoonual, Y., & Pongthanaisawan, J. (2013). Scenario analysis of electric vehicle technology penetration in Thailand: Comparisons of required electricity with power development plan and projections of fossil fuel and greenhouse gas reduction. *Energy Procedia, 34*, 459–470.

Tilburg, X.V., Röser, F., Hänsel, G., Cameron, L., & Escalante, D. (2012). *Status Report on Nationally Appropriate Mitigation Actions (NAMAs)-Mid-year update May 2012*. ECN and Ecofys. Retrieved from http://nama-database.org/images/f/fa/Annual_Status_Report_on_NAMAs_-_Update_May_2012.pdf

UNFCC. (2007, December). Decision-CP/13. Bali Action Plan. *Bali climate change conference*. Bali, Indonesia.

United Nations. (1992). *Report of the United Nations conference on environment and Development*. Rio de Janeiro. Retrieved from http://www.un.org/documents/ga/conf151/aconf15126-1annex1.htm

Universidad de los Andes. (2010a). *Curvas de costo de abatimiento de los Gases Efecto Invernadero- GEI y potenciales de mitigación en el sector industrial colombiano. Segundo Informe: Caracterización de la Industria Manufacturera, por sectores y sub-sectores*. Bogotá: Author.

Universidad de los Andes. (2010b). *Plan decenal de descontaminación del aire para Bogotá*. Bogotá, Colombia: Secretaría Distrital de Ambiente. ISBN: 978-980-6810-45-7. Retrieved from http://ambientebogota.gov.co/en/c/document_library/get_file?uuid=b5f3e23f-9c5f-40ef-912a-51a5822da320&groupId=55886

Universidad Nacional de Colombia – Sede Medellín. (2007). Evaluación de alternativas para la planificación energética sostenible de los sectores industrial y transporte del Área Metropolitana del Valle de Aburrá. Medellín.

US Energy Information Administration. (2011). *International Energy Outlook*.

Zapata, R.S. (2009). *Impacto en la demanda de energía eléctrica en Colombia debido a la penetración de vehículos híbridos-eléctricos y eléctricos*. Medellín: Universidad Nacional de Colombia – Sede Medellín.

Climate change mitigation actions in Brazil

Emilio Lèbre La Rovere, Amaro Olimpio Pereira Jr., Carolina Burle Schmidt Dubeux and William Wills

CENTROCLIMA/PPE/COPPE/UFRJ, Centro de Tecnologia, Cidade Universitária, Ilha do Fundão, CEP: Rio de Janeiro-RJ, Brazil

This article presents and discusses the main findings of a study focusing on climate change mitigation actions (MAs) in Brazil. Brazilian government presented to the 15th Conference of the Parties of the Climate Change Convention held in Copenhagen in 2009 (UNFCCC COP15) voluntary mitigation goals: a reduction in between 36.1 and 38.9% of the country's green house gases (GHG) emissions projected to 2020. If Brazil meets its voluntary goals, its GHG emissions in 2020 will be 6–10% lower than in 2005. The main research question addressed in this study is what conditions should Brazil meet to achieve its pledge? It discusses how MAs are approached, conceptualized and planned in Brazil. It describes how the country identifies MAs and the contextual framework used to define them. It also maps the initiatives underway in the country, and analyses the issues faced for its successful implementation, as well as for future approaches. The main conclusion is that Brazil seems to be in a good position to meet its voluntary mitigation goals up to 2020, as avoiding deforestation will take up the bulk of the emission reduction. After 2020, Brazil will face a new challenge: combine economic development with low GHG energy-related emissions.

1. Introduction

This article presents and discusses the main findings of a study developed for the Mitigation Action Plans and Scenarios (MAPS) programme[1] that relies on public documents and on a previous study on the Brazilian GHG abatement cost curves conducted by CentroClima (for details, see La Rovere et al., 2011).

Brazilian government presented voluntary mitigation goals to the 15th Conference of the Parties to the Climate Change Convention held in Copenhagen (UNFCCC COP15): a reduction in between 36.1 and 38.9% of the country's GHG emissions projected to 2020. This move was followed by other emerging economies, notably by China, India and South Africa. If Brazil meets its voluntary goals, emissions in 2020 will be 6–10% lower than in 2005. This reduction in absolute terms is far more ambitious than the pledges of other emerging economies. China and India have goals of reducing the GHG emissions intensity of GDP, but the high economic growth rates projected would imply a significant increase of absolute GHG emissions. South Africa's pledge is the reduction of GHG emissions compared to a baseline growth, but still implying an absolute increase of GHG emissions in 2020 (see Jiang, Hu, Liu, & Zhuang, 2009; MoEF, 2010; SBT, 2007).

The main research question addressed in this study is what conditions should Brazil meet to achieve its pledge? It discusses how MAs are approached, conceptualized and implemented in Brazil. It describes how the country identifies MAs and the contextual framework used to define them. It also maps the initiatives underway in the country and analyses the issues faced for its successful implementation, as well as for future approaches.

Section 2 provides an overview of Brazilian GHG emissions from 1990 to 2005, spotting the main GHG emission sources. Section 3 presents the Brazilian voluntary GHG emission mitigation goals introduced in Copenhagen (COP15) and confirmed in Cancun (COP16). The current status of mitigation actions (MAs) implementation in the country is described in Section 4. Finally, specific requirements for a successful implementation of the MAs for the major GHG emission sources (land use change, agriculture, forestry and animal husbandry; energy; and others) are discussed in Section 5. The final concluding remarks summarize the main results of this research study.

2. Brazil's GHG emission inventories

Brazilian second national communication to the UNFCC provides GHG emission inventories for the period from 1990 to 2005, presented in Table 1.

The main source of greenhouse gases in Brazil is deforestation caused by the expansion of agricultural frontiers, mainly in the Amazon region. Good estimates of deforested surface are available from satellite image recovery.

Table 1. GHG emissions in Brazil, 1990–2005.

GHG Emissions (M t CO_2eq/ano)	1990	1994	2000	2005	Var % 90/05	Share in1990 (%)	Share in 2005 (%)
Agriculture/Husbandry	347	378	401	487	41	25.4	22.1
Energy	215	256	328	362	68	15.8	16.4
Industrial processes	27	29	35	37	39	2.0	1.7
Wastes	28	32	41	49	77	2.0	2.2
Land use change	746	790	1247	1268	70	54.8	57.5
Total	1362	1485	2052	2203	62	100	100

Source: Brazil's Second National Communication, 2010.

However, the corresponding CO_2 emissions are very hard to quantify due to the lack of reliable values for biomass densities across the affected area.

Agriculture and husbandry are key sectors of Brazilian economy, and therefore rank second as main GHG emission sources. Brazil is one of the largest agricultural producers in the world, and ranks second in soybean production, with 18% of the global total. It also has the second largest bovine herd, with 12% of the global total. In this sector, CH_4 emissions are dominant, as a result of the enteric fermentation of ruminant herbivores, which include the huge cattle herd.

The energy sector comes only in the third position as GHG emitter, thanks to the role played by hydropower and renewable biomass (sugar cane ethanol, wood and charcoal from plantations, and biodiesel from vegetable oils cultivation), allowing for a 45% share of renewables in the country's total energy supply.

3. Brazil's voluntary GHG emission mitigation goals

Brazil has been making a lot of efforts to limit its GHG emissions, including curbing the Amazon deforestation and important investments on renewables. For the future, the National Climate Change Policy Law approved by the Congress and sanctioned by the President on 29 December 2009 (Federal Law n° 12187) included the voluntary goals to limit the country's GHG emissions presented the month before at COP15 in Copenhagen. The voluntary goals were established as a reduction in between 36.1 and 38.9% of the country's GHG emissions projected to 2020. Preliminary estimates of a business as usual (BAU) scenario[2] and of a mitigation scenario were made by several government bodies in the run to Copenhagen, discussed in the Brazilian Forum on Climate Change – FBMC, and constituted the basis of this pledge. These preliminary estimates are presented in Table 2.

However, the final figures had to wait for the completion of the Second National Communication in 2010. Therefore, it was only on 9 December 2010, during COP16 in Cancun, that Brazilian government published a decree (Federal Decree n° 7390) regulating the articles of Law n° 12187 regarding the final figures of the voluntary goals for the amount of avoided GHG emissions in 2020 as shown in Table 3.

Unlike the preliminary estimates made in 2009, the decree establishes the baselines, but not full mitigation scenario with voluntary goals for each main source of emissions. The exception is the energy sector, because the government considers the current 10-year energy plan as a mitigation scenario, as it includes a number of efforts to

Table 2. Preliminary estimates of Brazil's GHG emissions and mitigation actions in 2020.

Emissions/mitigation actions (M t CO_2eq/year)	2005 Inventory data	2020 BAU scenario	2020 Mitigation scenario	Reduction in 2020 (M t CO_2eq)	Reduction/BAU total in 2020 (%)
Land use change	1268	1084	415	669	24.7
Agriculture/Husbandry	487	627	461–494	133–166	4.9–6.1
Energy	362	901	694–735	166–207	6.1–7.7
Energy efficiency				12–15	0.4–0.6
Biofuels increase				48–60	1.8–2.2
Hydropower increase				79–99	2.9–3.7
Small Hydro, Biomass, Wind				26–33	1.0–1.2
Others	86	92	82–84	8–10	0.3–0.4
TOTAL	2203	2703	1652–1728	975–1052	36.1–38.9

Note: M t CO_2eq = million tons of CO_2eq.
Source: Brazils National Climate Change Policy Law (2009).

Table 3. Final figures of Brazil's GHG emissions and mitigation actions in 2020.

Emissions ($Mt\ CO_2eq/year$)	1990 Inventory data	2005 Inventory data	Variation 1990–2005 (%)	2020 BAU scenario	Variation 2005–2020 BAU (%)	Mitigation actions/ avoided emissions in 2020
Land use change	746	1268	70	1404	11	
Amazon				948		
Savannahs				323		
Others				133		
Agriculture/ husbandry	347	487	41	730	50	
Energy	215	362	68	868	140	234
Industrial processes + wastes	55	86	39	234	172	
Total	1362	2203	62	3236	47	1168–1259

Source: Brazil's Federal Decree n° 7390 (2010).

increase the role of renewables, nuclear and energy efficiency in the energy policy. MAs in the other sectors are only described but no corresponding emissions reductions are attributed. The main contribution to curb the country's GHG emissions will come from the efforts to reduce deforestation in the Amazon, following the successful record of recent years. The goal set for the agriculture sector is very ambitious, considering the recent growth of the country's grains and meat exports. However, economically feasible mitigation alternatives already exist and have a great potential: recovery of degraded pasture land, agroforestry schemes, more intensive cattle-raising activities (given the current low average ratio of 0.5 heads per hectare), biologic nitrogen fixation and low tillage techniques, which cover more than 20 million hectares in the country and are rapidly spreading.

In the case of the emissions of industrial processes and waste disposal, grouped under Other Sectors due to its minor contribution to the total, the BAU scenario already shows a low growth trend, and the voluntary commitments aim to keep roughly constant GHG emissions in these sectors. Again, there are feasible mitigation options in these sectors, such as the capture, burning and/or energy use of biogas in sanitary landfills that make it possible to achieve this objective.

The case of the energy sector deserves special attention. The emissions due to the use of fossil energy have been increasing significantly in the country from oil products, natural gas and coal. These fuels play a basic role to run the modern part of the Brazilian economy, such as industry and agribusiness, as well as transports, and the residential, commercial and service sectors. Their share in power generation has also been increasing, starting from a low departure level, to complement the use of the huge Brazilian hydropower potential, which is by far the dominant source. Thus, GHG emissions due to energy use, especially CO_2 from fossil fuels burning showed a high growth rate between 1990 and 2005, being 68% higher in the end of the period. Indeed, economic growth, the rising urbanization and the dominance of road transportation are the driving forces to increase fossil energy consumption and the associated CO_2 emissions.

Thus, unlike in other sectors, the BAU scenario projected by the government shows a significant increase in the emissions due to fossil fuel consumption by 2020, fostered by an average GDP growth projected at 5% p/y: a 140% rise compared to 2005, that is, 2.4 times the level of 2005 emissions in this sector.

As far as mitigation is concerned, the levels of hydropower generation, energy efficiency and alcohol production were those included in the 10-Year Energy Expansion Plan (PDE) for 2020 (EPE, 2010). Other mitigation actions included were the production and use of biodiesel in a 5% blend with diesel oil for 2020 (B5) and the increase in power generation from other renewable sources: small hydropower plants, biomass (especially sugarcane bagasse) and wind energy. Even so, GHG emissions from fossil fuels will be 75% higher in the mitigation scenario compared to 2005 emissions. The achievement of the mitigation scenario goals will require the implementation of public policy tools capable of stimulating renewable energy sources to replace fossil fuels. This need will be even more acute in the future to drive Brazilian economy towards a low carbon path, as fossil fuels will become the most important source of GHG emissions as elsewhere in the industrial world.

4. The implementation of Brazil's mitigation actions

4.1. *Legal instrument: Federal Decree n° 7390, 2010*

The Decree n° 7390 is a step towards detailing the voluntary mitigation goals established by the government in

Law n° 12187, but still leaves a lot of flexibility to the actual way of achieving these goals. Only the energy sector has already established the amount of avoided GHG emissions in 2020, according to the PDE. For the other sectors, the decree refers to four sectoral mitigation plans already elaborated:

- Plan of Action for Prevention and Control of Deforestation in the Amazon – PPCDAm
- Plan of Action for Prevention and Control of Deforestation and Fires in the Savannahs – PPCerrado
- Plan for Consolidation of a Low Carbon Emission Economy in Agriculture – ABC Plan
- Plan of Emission Reduction in the Steel sector

The decree lists a number of mitigation actions included in the available sectoral plans:

I. Reduction in 80% of the annual deforestation surface in the Amazon, compared to the historical average in the period 1996–2005; this figure is of 1.9535 M ha/year, and together with the average biomass density of 132.3 t C/ha (484 t CO_2/ha) was used to project the BAU emission level of 948 M t CO_2/y in 2020; assuming a constant biomass density, this decrease in the Amazon deforestation rate would allow for avoided emissions of 758 M t CO_2/y in 2020.
II. Reduction in 40% of the annual deforestation surface in the savannahs, compared to the historical average in the period 1999–2008; this figure is of 1.57 M ha/year, and together with the average biomass density of 56 t C/ha (206 t CO_2/ha) was used to project the BAU emission level of 323 M t CO_2/y in 2020; assuming a constant biomass density, this decrease in the savannahs deforestation rate would allow for avoided emissions of 129 M t CO_2/y in 2020.
III. Increase of renewable power generation through large hydropower, wind, small hydro and bioenergy projects, and of biofuels (ethanol from sugarcane and biodiesel from vegetable oils), and energy efficiency improvements as projected in the PDE; the amount of avoided emissions in 2020 was estimated at 234 M t CO_2/y considering that all this additional renewable energy generation and energy saved would come from fossil fuels.
IV. Recovery of 15 million hectares of degraded pasture land.
V. Increase in 4 million hectares of the land covered by agroforestry schemes, coupled with more intensive cattle raising activities (integrated agriculture/husbandry/forestry activities).
VI. Increase in 8 million hectares of the planted area under low tillage techniques.
VII. Increase in 5.5 million hectares of areas cultivated with biologic nitrogen fixation techniques replacing the use of nitrogenous fertilizers.
VIII. Increase in 3 million hectares of forest plantations.
IX. Increase in 4.4 million cubic meters the use of technologies for proper treatment of animal wastes.
X. Increase steel manufacturing using charcoal from planted forests and improve the efficiency of charcoal kilns.

4.2. *Additional sectoral plans to achieve the mitigation actions*

The Decree also establishes a requirement for the elaboration of additional sectoral mitigation plans for those sectors included in the 2009 law:

- Public urban transportation
- Interstate transport of cargo and passengers
- Transformation industry
- Durable consumer goods industry
- Chemical industry
- Pulp and paper industry
- Mining
- Civil construction
- Health sector

All these sectoral mitigation plans must include:

- Emission reductions in 2020, with milestones for every 3-year period;
- Mitigation actions to be implemented;
- Establishment of indicators for monitoring of performance and assessment of effectiveness;
- Proposal of tools and incentives to be adopted in the implementation of the plans;
- Sectoral studies of cost estimates and implications for competitiveness.

The Decree also requires a wide public consultation process for the discussion of the sectoral mitigation plans, and allows for the possibility of using the sectoral mitigation goals in the establishment of the Domestic Carbon Market authorized by the 2009 law. The institutional responsibility of coordinating the actions of the sectoral mitigation plans, under the umbrella of the National Climate Change Plan, remains with the Interministerial Commission on Global Climate Change – CIMGC, while the Brazilian Climate Change Forum – FBMC is responsible for the follow-up of the actions implementation. A working group coordinated by the Ministry of Science and Technology – MCT will be responsible for publishing an annual estimate of the country's GHG emissions.

The Decree also requires that all federal government multiyear plans and annual budget laws include the provision for the mitigation programmes and actions included in the decree.

4.3. *Financial incentive mechanisms*[3]

Other tools to implement the sectoral mitigation plans include the Clean Development Mechanism[4] (CDM) projects, carbon markets and the 'nationally appropriate mitigation actions' (NAMAs), according to the decree, besides the National Climate Change Fund created in 2009 by Federal Law n° 12114.

Brazil has existing sources of funding for energy efficiency and renewable energy through government-mandated levies; these are directed towards funds such as the Fuel Consumption Account (CCC) for power off grid generation in the Amazon region, Energy Development Account (CDE), and Global Reversion Reserve (RGR). According to the National Agency for Electric Energy (ANEEL), CDE collections for 2009 are estimated at €1.12 billion. Collections for the RGR Fund had reached nearly €3 billion at the end of fiscal year 2008. Managed by Eletrobrás, RGR is a main source of funding for energy-efficiency programmes under the Electricity Conservation Program (Procel). With regard to the CCC, collected levies totaled approximately €0.5 billion in 2008. Not all of the funds collected are used for renewable energy or energy-efficiency projects, but they are significant in size.

Other funds that receive similarly mandated levies but are not limited to energy-related activities include the Constitutional Financing Fund of the Northern, Northeastern and Center-West Regions (FNO, FNE, and FCO, respectively). These funds receive 3% of the overall tax collections, which are then used to finance activities in the respective regions, managed by official banks, with an overall budget of €5.25 billion in 2009. Financing programmes include support for activities such as decreased deforestation and increased livestock productivity.[5]

Additionally, to financial instruments in place, Seroa da Motta (2011) proposes that the Domestic Carbon Market could profit from the low mitigation cost related to avoided deforestation.[6]

5. Specific issues related to the implementation of Brazilian MAs

The successful achievement of the country's voluntary GHG emission mitigation goals will mainly depend on the capacity to handle a number of issues that arise in the implementation of MAs, including (La Rovere, 2011)

- Development of these mitigation actions (from the idea to concept note, business plan and successful implementation);
- Planning and establishing policy and regulatory context[7] (both are of individual mitigation actions and broader plans and strategies);
- Setting up institutional capacity to take mitigation actions to implementation;
- Creating technical capacity to design and domestically assure the monitoring, reporting and verification (MRV) of the mitigation actions;
- Scheduling the means to financing these mitigation actions;
- Outlining the ownership of the mitigation actions; and
- Certifying the credibility of MAs MRV.

These issues are briefly discussed below for the main sources of GHG emissions.

5.1. *Land use change*

This is the most important source of GHG emissions, and successful efforts to reduce deforestation particularly in the Amazon and 'Cerrado' regions will be crucial, as shown in Sections 3 and 4. As mentioned in Section 4.1, two Plans, gather the MAs defined to meet the voluntary goals: PPCDAm for the Amazon region and PPCerrado for the Cerrado region.

The Ministry of Environment – MMA, takes the lead in the definition and implementation of these MAs, coordinating the action of other ministries, such as MCT, the Ministry of Defense and the Ministry of Justice, and of subnational governments (states and municipalities). The key target is the enforcement of laws and regulations that prevents illegal deforestation of conservation units and private properties. The Forest Code is the most important piece of legislation in this regard, as it rules that only 20% of the original forest coverage of private properties can be cleared in the Amazon region.

The operational bodies within MMA (The Forest Service and the National Environmental Agency – IBAMA) have the technical capacity to design the modalities of implementing these MAs. However, the institutional capacity to actually enforce the laws and regulations, particularly in the Amazon region, is insufficient to cope with the powerful economic driving forces of deforestation (the dynamics of expansion of the agricultural frontier, led by logging, cattle raising and soybean plantations). A strong political disposal is required to keep all the governmental institutions mobilized to constantly verify the respect of the law.[8] Recent records have shown the viability of a successful performance in this field, as deforestation in the Amazon was limited to 1.2 M ha in 2007 and 0.7 M ha in 2008, 0.27 M ha in 2011 and 0.205 M ha in 2012 down from an average of 1.95 M ha/year in the period 1996–2005.

Owing to sovereignty concerns, all the financing of these MAs will come from the national budgets at the federal, state and municipal levels.

The main MRV issue related to these two plans is the estimate of the biomass density of the forest cover in the deforested surface, as satellite imagery provides a reasonable guess of the affected area (still, some small scale deforested areas may be difficult to spot). As a forest inventory is not available for the whole Amazon, a special methodology was developed for calculating the GHG emissions inventory of the country, from 1990 to 2005, in Brazil's Second National Communication to the UNFCCC (2010), improving the work done in the First National Communication (2004). This methodology was developed by the National Spatial Research Institute – INPE, belonging to MCT. An independent monitoring of the deforestation in the Amazon is also carried out by a NGO, the Amazon Research Institute – IPAM. Therefore, MRV of these MAs will have the same credibility of Brazilian National Communications to the UNFCCC.

Besides these two Plans, it is worth mentioning a different kind of effort that might turn out to develop into MAs in the medium or long-term future: the Amazon Fund, established in 2009. Funding in this case comes from international cooperation, through a pioneer 1 billion US$ grant from the Norwegian government, and will also benefit from other grants from international donors (the German government and several NGOs have announced their support to the fund). However, all the decisions concerning the allocation of funds are solely the responsibility of the National Economic and Social Development Bank – BNDES, a key condition that had to be met in the set up of the Fund to dismiss any sovereignty concern. In its first year of operation, the Fund has already received 45 proposals and has started supporting five projects implemented by NGOs in the Brazilian Amazon. BNDES is preparing to extend the coverage of the Fund to support projects in neighbouring countries in the Amazon region. Most of the projects are initiatives from NGOs related to innovative approaches to supply economic incentives to keeping the forest cover in the Amazon, including the reduction of emissions from deforestation and forest degradation – REDD+. They cannot be labelled as MAs today, as GHG-avoided emissions are not the main deliverable of these projects, but these experiments might well develop into MAs in the future.

The contrast between the Amazon and 'Cerrado' Plans, on one hand, and the Amazon Fund, on the other, allows illustrating the borderline drawn by Brazilian government in the delicate issue of sovereignty × international support to MAs. The enforcement of laws and regulations remains the sole responsibility of the government, without any financial support from abroad. However, it is acknowledged that driving forces of deforestation are very powerful and demand the establishment of economic incentives to keep the forest cover in the long term, going far beyond the sole command and control policies and measures. International cooperation towards this end is most welcome.

5.2. *Agriculture, forestry and animal husbandry*

This is currently the second most important source of GHG emissions in the country. The Plan for Consolidation of a Low Carbon Emission Economy in Agriculture (ABC Plan) gathers the MAs defined to meet the voluntary goals mentioned in Section 4.1

The Ministry of Agriculture takes the lead in the definition and implementation of these MAs, coordinating the action of other ministries, such the Ministry of Planning, Budget and Management and the Ministry of Economy and of subnational governments (states and municipalities). The key policy tool is the establishment of eligibility requirements for farmers to get credit from governmental development banks, and of economic incentives to access softer loans from these public bodies (mainly from the Banco do Brasil but also from BNDES and others). This is already the current practice of these players (e.g. farming requirements from zoning plans must be met to access public funds) that may be extended to integrate MAs.

The financing of these MAs will come from the usual sources of funding to national development banks, including the financial market, besides national budgets at the federal, state and municipal levels. It is thus conceivable that in the future these MAs may become candidates for financial support to NAMAs to be provided through UNFCCC mechanisms. As the targets can be considered very ambitious, this financial support could be valuable to improve the institutional capacity in the country to meet them. The Brazilian development banks that are the primary source of credit to agricultural, forestry and animal husbandry activities would be potential owners of the MAs under the political coordination of the Ministry of Agriculture.

The operational body dealing with research and development within MA, the Brazilian Agriculture Research Enterprise – EMBRAPA, a network of research centres, has the technical capacity to design the modalities of implementing these MAs and also to deal with the MRV-related issues.

5.3. *Energy*

As mentioned, emissions due to the use of fossil energy have been increasing significantly in the country and the most recent 10-year Energy Plan – PDE, covering the period 2011–2020 (EPE, 2010), is already considered by the government as a sectoral mitigation plan.

MME is responsible for approving and implementing the country's energy policy. The technical inputs to MME are provided by the Energy Planning Agency – EPE, which formulates the studies leading to the PDE, detailing the energy programmes and projects to be implemented in the next 10 years period. The listed power plants are then offered to the private sector in several rounds of call for tenders to build them, and a similar process is run for oil & gas fields.

The actual investment to implement these MAs will thus come from the private sector or state-owned enterprises such as Petrobras and Eletrobras. In the case of biofuels, and particularly of ethanol production from sugarcane in the Brazilian context, the profitability of these investments in the current oil price[9] scenarios leads to the assumption that there will be no major problems in the flow of private funding to these MAs.

The cases of energy efficiency and of renewable power generation are different due to the well-known barriers that hamper their development. Economic incentives to promote renewables are already supplied from Brazilian federal budget (mainly through BNDES and Eletrobras) and also strengthened by some punctual initiatives at the state and municipal levels. However, these MAs would be the most natural candidates to Brazilian NAMAs seeking financial support from the UNFCCC funding mechanisms. Other possibilities of international cooperation would be the multilateral financial bodies such as the World Bank and IDB, which may resume the role they had in the past for hydropower development, now under updated standard of environmental and social performance (such as those established by the World Commission on Dams, for example). Again, as PDE targets can be considered ambitious, this financial support could be valuable to improve the institutional capacity in the country to meet them.

Public players in this field, such as EPE, Eletrobras and BNDES, would be potential owners of the MAs, under the political coordination of MME.

EPE has the technical capacity to design the modalities of implementing these MAs and also to deal with the MRV-related issues, which are pretty straightforward to solve, being much less complex than for other GHG emission sources. The key issue will remain the inherent subjectivity associated to the choice of a counterfactual baseline to be compared with the actual energy policy followed by the country. There is no scientific approach capable of solving this issue that must be settled in the international negotiations table.

5.4. *Others*

For the bulk of other MAs, which are less significant in terms of total GHG-avoided emissions, the Sectoral Mitigation Plans are under development and are expected to be approved in the end of 2012 (initially scheduled for December 2011, and suffering from repeated delays). The sole exception is the already available Plan of Emission Reduction in the Steel sector, including MAs targeted to increase steel manufacturing using charcoal from planted forests and improving the efficiency of charcoal kilns. This case combines issues from forestry and energy efficiency measures.

The Mining Sector Mitigation Plan is based on the Emissions Inventory sector carried out by the Brazilian Mining Institute (IBRAM) for 2008 and on the 2030 National Mining Plan. In the Mitigation Plan, sectoral emissions are estimated to be around 17.6 M t CO_2e in 2020, with a potential reduction of 4%. The plan is focused on the mining activity itself, not considering external transportation and processing. The activities covered by the projections represent 80% of national production, include 14 mineral types and are mainly focused on the iron, copper, aluminum and coal production.

The Sectoral Mitigation Plans for the Industry and Transport sectors are the most important ones, due to the relevance of the GHG emissions, and will need to be consistent with the estimates done by the Energy Planning Agency in the 10-year and Long-term (2040) National Energy Plans. The Transport Plan is detailing the investments required to expand the ethanol penetration up to 75% on the light vehicles fleet, the investments on the infrastructure required to absorb the fast-growing light vehicles fleet in Brazil and the modal shifts potential provided by investments in public transport.

The key challenges for MRVs of NAMAs remain within the Transport sector, particularly at the subnational level (cities and states). For the other mitigation actions (energy efficiency, renewable power generation, methane recovery and use from urban solid wastes), CDM methodologies are available and are straightforward to be applied at a programme level. However, the estimate of avoided GHG emissions by mitigation actions on intracity passenger transportation, such as Bus Rapid Tranportation Systems – BRTs, subway lines, light rail and improvement of the bus system (rerouting, integrated ticketing, exclusive bus lanes) is much more complex as it requires the proper accounting of modal shifts. This is data intensive, requires primary data surveys and is very site specific. Moreover, cities do not always have the appropriate institutional setting (public transportation agencies keeping track of transport data with effective performance). The city of Rio de Janeiro alone is building five BRTs and extending subway lines, and the only estimates available for the avoided GHG emissions come from rough analogies with the Transmillenium project, using its CDM methodology (La Rovere, Carloni, & Buzzatti, 2012). The same applies to São Paulo and Brasília. The sole exception is Belo Horizonte, where a good transportation municipal agency with support from IDB has recently provided a tailor-made estimate of the GHG emissions to be avoided through its Municipal Transportation Plan.

6. Conclusions

The successful achievement of the Country's voluntary GHG emission mitigation goals will depend on the internal capacity to handle a number of issues that arise in the implementation of MAs, including (La Rovere, 2011):

- Development of these mitigation actions (from the idea to concept note, business plan and successful implementation);
- Planning and establishing policy and regulatory context (both of individual mitigation actions and broader plans and strategies);
- Setting up institutional capacity to take mitigation actions to implementation;
- Creating technical capacity to design and domestically assure the MRV of the MAs;
- Scheduling the means to financing these mitigation actions;
- Outlining the ownership of the mitigation actions; and
- Certifying the credibility of MAs MRV.

Moreover, accomplishments on overlaps and gaps regarding the institutional and regulatory framework should allow better governance conditions to the Country's climate change policy. Furthermore, as presented in the paper, the challenges related to the implementation of the Brazilian NAMAs are actually tough. The specificity of the Country's GHG emissions structure and dynamics, the high level of renewables in the internal energy mix and the perspective of a huge intensification of oil & gas production from the pre-salt offshore fields are major issues that should be taken in account in a mid- to long-term perspective of climate change policy. These challenges will be increasingly important after 2020, when the focus of the mitigation actions will necessarily shift from the land use change to the energy system.

As land use change has been the most important source of GHG emissions in the country, successful efforts to reduce deforestation particularly in the Amazon and 'Cerrado' regions will be crucial to the achievement of Brazilian voluntary mitigation goals. The key target is the enforcement of laws and regulations that prevent illegal deforestation of conservation units and private properties. The recent record has shown the viability of a successful performance in this field, as deforestation in the Amazon was limited to 0.205 M ha in 2012, down from an average of 19.5 thousand km^2/year in the period 1996–2005. Therefore, Brazil seems to be in a good position to meet the voluntary mitigation goals pledged to the UNFCC up to 2020, as avoided deforestation will take up the bulk of the emissions reduction. In 2020, if governmental mitigation goals are met, GHG emissions from the energy system will take over and become the largest in the country.

After 2020, Brazil will be in a situation more similar to other industrialized countries, faced with a new challenge of economic development with low GHG energy-related emissions. If no additional mitigation policies and measures are implemented, GHG emissions will start to increase again in the period 2020–2030, due to population and economic growth driving energy demand, supply and GHG emissions up. In order to avoid this, the portfolio of additional mitigation actions would have to be substantially extended.

On the other hand, the challenge to curb down GHG emissions in Brazil is very different of the one faced by other emerging economies such as China, India and South Africa. These countries heavily rely on coal to supply their energy needs and thus have much higher emission intensities per GDP and mitigation potentials than Brazil (see Jiang et al., 2009; MoEF, 2010; SBT, 2007). Yet, the ambition of their voluntary goals is much lower than in the Brazilian case due to the time required to develop and deploy low-carbon energy technologies.

However, Brazil is in a privileged position to take a lead in low-carbon economic and social development due to its huge endowment of renewable energy resources. On the demand side, it seems imperative to limit oil consumption in the transport sector, heavily dominated by road transport of cargo and cars in cities. Modal shifts towards railways and waterways, as well as the long due development of mass transportation infrastructure in cities, will be required. Unlike the mitigation actions in the land use change sector, where most of the funding will come from the national budgets, due to sovereignty concerns, the huge financial resources needed to develop low-carbon transport infrastructure could benefit from soft loans channelled to the country through NAMAs (La Rovere, 2011; La Rovere & Raubenheimer, 2011).

Notes

1. The MAPS programme is a collaboration between a number of developing countries, promoting best practice in mitigation action planning and scenarios development, by supporting in-country processes informed by research. It seeks to share and deepen knowledge through collaboration between Southern experts supporting government programmes.
2. It is worth highlighting, as mentioned before, that if the goals are met, GHG emissions in 2020 will be 6–10% lower than in 2005, regardless the baseline pathway.
3. This section is extracted from La Rovere and Poppe (2012)
4. For details see Gutierrez (2011).
5. Total funding needs to implement all MAs are still to be estimated after the completion of the sectoral plans currently under discussion. For a mitigation cost assessment of the Brazilian NAMAs see La Rovere et al. (2011) and for general cost curves for the country, see De Gouvello et al. (2010).
6. For details on avoided deforestation costs, see Margulis, Dubeux, and Marcovitch (2011) and Seroa da Motta (2005).

7. For details on governance, see Seroa da Motta (2011).
8. For a full discussion on the reduction of deforestation rates in Brazil see Assunção, Gandour, and Rocha (2012).
9. For a detailed discussion of ethanol competitiveness and oil prices, see Cavalcanti, Szklo, Machado, and Arouca (2012).

References

Assunção, J., Gandour, C., & Rocha, R. (2012). *Deforestation slowdown in the legal Amazon: Prices or policies?* Rio de Janeiro: Climate Policy Initiative/PUC-Rio.

Brazils Federal Decree n° 7390 sanctioned by the President on December 9, 2010.

Brazil's Initial National Communication to the United Nations Framework Convention on Climate Change. (2004). Ministry of Science and Technology. Brasília, November 2004.

Brazils National Climate Change Policy Law approved by the Congress and sanctioned by the President on December 29, 2009 (Federal Law n° 12187).

Brazils National Climate Change Fund Law approved by the Congress and sanctioned by the President on December 9, 2009 (Federal Law n° 12114).

Brazil's Second National Communication to the United Nations Framework Convention on Climate Change. (2010). Ministry of Science and Technology. Brasília, November 2010.

Cavalcanti, M., Szklo, A., Machado, G., & Arouca, M. (2012). Taxation of automobile fuels in Brazil: Does ethanol need tax incentives to be competitive and if so, to what extent can they be justified by the balance of GHG emissions? *Renewable Energy, 37,* 9–18.

De Gouvello, C., Soares Filho, B.S., Nassar, A., Shaeffer, R., Alves, F.J, Alves, J.W.S., & La Rovere, E.L. (2010). *Brazilian low carbon country study.* Washington, DC: The World Bank.

EPE. (2010). Plano Decenal de Expansão – PDE 2011–2020.

Gutierrez, M.B.S. (2011). From CDM to nationally appropriate mitigation actions: Financing prospects for the Brazilian sustainable development. In R. Seroa da Motta, Jorge Hargrave, Gustavo Luedemann, & Maria Bernadete Sarmiento Gutierrez (Eds.), *Climate Change in Brazil: Economic, social and regulatory aspects* (pp. 139–156). Brasília: IPEA.

Jiang, K., Hu, X., Liu, Q., & Zhuang, X. (2009). *Low-carbon economy scenario studies up to 2050.* Beijing: Energy Research Institute, National Development and Reform Commission.

La Rovere, E.L. (2011). *Mitigation actions in Brazil. Country Study for the mitigaton action plans and scenarios.* Rio de Janeiro: MAPS Programme.

La Rovere, E.L., Carloni, F.B., & Buzzatti, M. (2012). Proposal of a Monitoring System of GHG Emissions Reductions for the City of Rio de Janeiro, Government of the City of Rio de Janeiro, Environment Secretariat/CentroClima/COPPE/UFRJ, March 2012.

La Rovere, E.L., Dubeux, C.B.S., Pereira, Jr. A.O, Medeiros, A., Carloni, F.B., Turano, P., & Fiorini, A.C. (2011). *GHG emissions scenarios for Brazil and a cost-efficiency analysis of mitigation actions.* Rio de Janeiro/Brasília: CentroClima/PPE/COPPE/UFRJ for the Brazilian Ministry of Environment – MMA and UNDP – United Nations Development Program, Project BRA/00/020. Support to Public Policies in Management and Environmental Control, Summary Report, July 2011, 96 p. Retrieved from http://www.lima.coppe.ufrj.br/includes/pages/clipping/PNUD_Produto_5.1.pdf

La Rovere, E.L., & Poppe, M. (2012). *Brazilian climate change policy.* Paris: Institute for Sustainable Development and International Relations – IDDRI, February 2012, 31 p.

La Rovere, E.L., & Raubenheimer, S. (2011). Low carbon scenarios in emergent economies: The Brazilian case, Low Carbon Society Research Network 3rd Meeting, Paris, 13–14 October 2011, 3 p.

Margulis, S., Dubeux, C.B.S., & Marcovitch, J. (2011). Redução do Desmatamento na Amazônia e seus Custos de Oportunidade. In S. Margulis, C. Dubeux, & J. Marcovitch (Eds.), *Economia da Mudança do Clima no Brasil.* Rio de Janeiro: Synergia.

MoEF (Ministry of Environment and Forests, Government of India). (2010). *India: Taking on climate change post-Copenhagen domestic actions.* New Delhi: Ministry of Environment and Forests, Government of India. Retrieved from http://moef.nic.in/downloads/public-information/India Taking on Climate Change.pdf

SBT (Scenario Building Team). (2007). *Long term mitigation scenarios: Strategic Options for South Africa.* Pretoria: Department of Environment Affairs and Tourism. Retrieved from http://www.environment.gov.za/HotIssues/2008/LTMS/A LTMS Scenarios for SA.pdf

Seroa da Motta, R. (2005). Custos e Benefícios do Desmatamento na Amazônia. Ciência & Meio Ambiente, v.32.

Seroa da Motta, R. (2011). The National policy on climate change: Regulatory and governance aspects. In R. Seroa da Motta, Jorge Hargrave, Gustavo Luedemann, & Maria Bernadete Sarmiento Gutierrez (Eds.), *Climate change in Brazil: Economic, social and regulatory aspects* (pp. 33–44). Brasília: IPEA.

A case study of Chilean mitigation actions

José Eduardo Sanhueza and Felipe Andrés Ladron de Guevara

Climate Change and Development Consultants, Santiago, Chile

The Bali Action Plan established that the design of a future international agreement must consider enhancing the developing countries participation in greenhouse gas mitigation activities in order to achieve the ultimate objective of the United Nations Framework Convention on Climate Change. Such participation should be appropriate to national circumstances and supported technologically and financially by the industrialized nations. This decision has generated particular interest because it implies a change in the way developing countries have until now faced their differentiated responsibility, with qualitative commitments to mitigation but not quantified obligations. In 2010, Chile communicated that it will achieve a 20% reduction below the 'business as usual' emissions growth trajectory in 2020. This paper first describes the progressive involvement of Chile in mitigation actions tasks with a periodization that shows the evolution of taking on obligations and opportunities within the framework of the Convention. In its final part, the paper elaborates on the last stage of this process and briefly presents progress as of mid-2012 on identifying Chile's (potential) Nationally Appropriate Mitigation Actions in the transport, agriculture and energy sectors. The assessment of the proposals of the more advanced cases analyzed concludes that the feasibility of realizing and achieving their objectives not only will depend on the availability of financial resources, but also, and primarily, on the political will of the Government.

1. Introduction

One of the fundamental principles underpinning the United Nations Framework Convention on Climate Change (UNFCCC) in 1992 is that of 'common but differentiated responsibilities and respective capabilities' (UNFCCC, 1992). This meant at the time that, *inter alia*, developed countries were exclusively responsible for taking mitigation actions to address the problem of climate change. Developing countries could voluntarily contribute to these efforts, to the extent that they were supported financially by developed countries.

Therefore that for many years the participation by the developing world in climate change mitigation actions has been mainly limited to implementation of mitigation activities financed by the Global Environment Facility (GEF) and later to actions implemented by private bodies and facilitated by the Clean Development Mechanism (CDM) of the Kyoto Protocol.

A significant change was made at the Conference of the Parties of the UNFCCC at its 13th gathering (COP13), which took place in Bali, Indonesia in 2007. The situation for developing nations shifted at these negotiations on the necessary elements for reaching an agreement on long-term international co-operation towards achieving the goals of the convention. Developing countries agreed 'to consider, *inter alia*, nationally appropriate mitigation actions in the context of sustainable development, supported and enabled by technology, financing and capacity-building, in a measurable, reportable and verifiable manner' (UNFCCC, 2007).

So, at the 2007 conference, the principle of common but differentiated responsibilities and respective responsibilities was 'reunderstood', and since then there has been a considerable increase in the number of proposals for mitigation actions in the developing world that seek internationally pledged support for quantifiable mitigation actions that reduce emissions growth. In 2010, Chile communicated that it will pursue a 20% reduction below the 'business as usual' emissions growth trajectory in 2020, with projections starting from the year 2007 (Chile, 2010).

Chile's climate change policy has not been an exception to the above as is illustrated in the following sections outlining Chile's participation in mitigation actions in this scenario and its evolution until the Nationally Appropriate Mitigation Actions (NAMA) stage.[1] The focus of this article is not on the overall pledge, but the development of NAMAs informed by this broader context. The focus of this paper is on the evaluation of Chile's engagement on climate change, across four periods, in meeting its obligations under the Convention and making use of opportunities provided.

2. Approach to mitigation actions in Chile

Chile's involvement in addressing the challenge of climate change – a problem caused by current patterns of development – has been no different to those of most developing nations. They have been responses to international agreements signed on this matter: the UNFCCC and the Kyoto Protocol.

The country's participation in addressing the problem of climate change began upon signing the Framework Convention in 1992 at the summit in Rio, a commitment that was ratified by the Chilean Congress in 1994. Since then, specific phases of development in Chile's involvement in climate change tasks, and more specifically mitigation action, can be distinguished.

2.1 *The first phase (1992–2001)*

The first phase extending between the year of signing the convention in 1992 and 2001 and was characterized by: (a) thc initial usc thc GEF, thc operating entity of the Convention's financial mechanism, enabling developing countries to fulfil their commitments, and (b) the development of an institution for creating a country position on the issues discussed at the convention.

In 1992, only a few months after signing the convention and during the pilot phase of the Facility, Chile received funding from the GEF for two projects. The first project dealt with promoting installation of services supporting efficient use of energy, using Energy Service Companies, through pilot projects in the copper mining sector in the country. The second project consisted of a study of the economic feasibility of producing methanol from lignocellulosic material, which is abundant in Chile, and which could have a great impact on CO_2 emissions originating in the transport sector (Global Environmental Facility, 1992).

Unfortunately, a lack of integration of these projects into national policy or programmes resulted in their isolation, administrative difficulties with their implementation could not be overcome, being implemented by international consultants resulted in lack of local ownership, and the projects were not able to reach their goals.

A second proposal for mitigation actions to the GEF in 1999 was more successful. The project was titled 'Removal of obstacles to electrification of rural areas using renewable energies' (Global Environmantal Facility, 1999). This project proposed aligning goals of the Convention with a successful policy of rural electrification, which the government had already been implementing in the 1990s and applied for funding from the fund to cover the incremental costs of the policy using renewable energy sources. The Project was implemented in the first part of 2000 and most of the goals were reached.

In 1996, the country applied to the GEF for resources in order to fulfil its commitment to preparing its First National Communication for the convention (Global Environmental Facility, 1996). In this, the country presented an inventory of greenhouse gas (GHG) emissions in Chile in 1994, a first GHG emission forecast for 2020, and possible scenarios, with some mitigation actions, which could curb the current emission growth trend.

It must be mentioned that in the framework of this communication, a calculation of national emissions of GHG was also made, from both the energy sector and non-energy sector, for the period between 1984 and 1998. This historic information is very valuable for mapping out trends and has been periodically updated since.

Considering that the issue of climate change was becoming increasingly relevant to the country, both in terms of international negotiation processes and initiation of cooperative projects, Chile decided to set up an institutional authority to facilitate debate and government consultation in decision-making., A National Advisory Committee on Global Change was set up in 1996, with representatives from the National Environmental Commission, the Ministry of Foreign Affairs, the Ministry of Agriculture, the National Energy Commission, the General Directorate of Maritime Territory and Merchant Marine, the Meteorological Directorate of Chile, the Hydrographic and Oceanographic Service of the Chilean Navy, the National Commission for Scientific and Technological Research and the Chilean Academy of Sciences.

The committee only began to operate on a regular basis from the beginnings of 1998, and in a short time opened spaces for more active participation within the country in discussions concerning modalities and procedures of the Kyoto Protocol, particularly in that of the CDM (UNFCCC, 1997), that consequently allowed for a more active participation in the international negotiations on these issues.

Focussing on the medium-term, another central task of the Advisory Committee was to develop strategic guidelines for climate change in Chile, with the aim of establishing a framework for government action in order to realize the commitments under the Convention. These guidelines were approved by the Board of the National Environment Commission in 1998.

2.2 *Second phase (2002–2005)*

The second distinguishable phase of actions dealing with climate change dates between 2002 and 2005 and is characterized by extensive use of the CDM. This period began with a study on a strategic use of this market-based instrument (Sanhueza Flores, 2003), followed by the establishment of the Designated National Authority (DNA) for the CDM in Chile and an intense campaign promoting opportunities for implementing CDM projects. The national agency which promotes national products in international markets (PROCHILE) participated actively in this campaign. Also, campaigning was done at the Carbon Fairs.

These efforts were successful in that Chile, in those years, was ranked the country with the most registered CDM projects. In addition, important human and institutional capacity was developed to support preparation of CDM projects, according to CDM modalities and procedures which were broadly agreed in Marrakech in 2001.

It is important to note that a number of conditions contributed to this situation. First investors in the emerging carbon market had confidence in Chile because of its political stability, the security offered for its legislation on foreign investment and the simplicity of the procedures set out by the DNA for endorsing CDM proposals. The latter was reduced to providing proof of an environmental permit issued for the project, which is already required by national legislation before implementation begins.

Attempts were made by Chilean leadership to integrate these developments into national policies, mainly in the energy sector. This was done, for example, by issuing subsidies for studies such as a technical feasibility study of renewable energy initiatives and a pilot project to promote the use of the instrument in both small and medium national industry. Yet despite these efforts, progress in integrating CDM projects into national policy was marginal.

In 2001, the GEF approved two of the country's new projects on climate change mitigation. One of them was a small scale project to build capacity for assessing technological needs and promoting integration of climate change concerns into development planning and goal setting (Global Environmental Facility, 2001). The other project consisted of carrying out a series of studies to promote the reduction of GHG emissions in land transport in Santiago, which involved promoting the use of bicycles, modernizing the public bus system and improving traffic (Global Environmental Facility, 2003). These later studies, only a few of many others carried out in Chile on this issue, recently informed the development of a rapid-bus system in the capital, known as the Transantiago, and also the important development of a network of cycle-lanes.

The period ended in international terms with the entry into force of the Kyoto Protocol in 2005. In Montreal, discussions were also launched to broaden participation to include the USA and developing countries.

2.3 *Third phase (2006–2009)*

The third phase dates between 2006 and 2009, leading up to COP-15 in Copenhagen (2009), which sought to negotiate a new climate treaty. The period was characterized by extensive work on laying the foundations for national awareness on the climate change problem; to providing climate change policy; and to establishing a procedure for structuring and implementing this policy.

The main domestic reasons for readdressing the issue of climate change at these years included

(1) a difficult situation due to unreliable supply of energy, exacerbated by periods of pronounced droughts and problems with accessing natural gas from neighbouring countries which highlighted the extreme fragility of Chile's energy system,
(2) publication of a study of climate variability in Chile in the twenty-first century. The study used the PRECIS model, developed by the Meteorological Office in the UK, and illustrated in detail the effects that uncontrolled increase in global temperature could have on the economy and population, and
(3) Chile's intention to become a member of the Organisation for Economic Co-operation and Development (OECD) and the guidelines set out for members of the organization.

The unreliability of energy supply in the country, accompanied by rapid economic growth, saw increased efforts of securing a more efficient energy supply and to using alternative and renewable sources. In order to achieve this, the basis was established in this period for development of two appropriate policies.

Firstly, the Country Energy Efficiency Programme was instituted at the beginning of 2005, under the wings of the Ministry of Economy and Development and Reconstruction. However, it only began to operate from the 1 December 2008 when administration was handed over to the Chilean Agency for Energy Efficiency, an implementation body of the Ministry of Energy which integrates the goal of energy efficiency across national productive activities.

The second policy encourages the use of renewable energy sources available in the country. The government has taken a two pronged approach to achieving this; legislative and supportive components.

The new law approved in Chile, promoting non-conventional renewable energies (NCREs) (Law No. 20.257), came into practice in 2010 and states that energy generating companies in the country, with a capacity of 200 MW or more, must ensure that 10% of their energy generation each year originates from NCRE sources. These can be either owned by the company or contracted in.

This legislation will be phased in with the required quota starting at 5% between 2010 and 2014, with an increase of 0.5% in 2015 and the final quota of 10% set for 2024. Electrical energy distributors supplying regular consumers are required to reach the 10% quota by 2010.

Simultaneous to processing and promulgating the legislation, the Government established the Centre for Renewable Energies. The aim is to promote and facilitate the development of the NCRE industry and to coordinate both public and private efforts that optimize the vast potential of NCREs existing in Chile.

In this third period, the country submitted two proposals to the GEF in support of these policies. One proposal was to promote and strengthen the energy efficiency market in the country's industrial sector, and was approved in 2010 (Global Environmental Facility, 2010). The second proposal focussed on promoting and developing solar technologies for heating water and generating electricity, and was approved in 2009 (Global Environmental Facility, 2009).

Parallel to these developments, there were also others in this period. The first of them began in 2006, when the Council of Ministers of the National Environment Commission (CONAMA), the environmental institution that existed prior to the current Ministry of Environment, was entrusted with preparing the first 'National Action Plan on Climate Change' (Ministry of Environment, 2008). This was completed at the end of 2008, and lays out plans for the period between 2008 and 2012.

This Action Plan is a frame of reference for assessment activities of the impacts of climate change and vulnerability and adaptation to climate change, as well as mitigation of GHG emissions in the country. It was drawn up through a process of consultation with participants such as technical personnel from institutions, part of the CONAMA Council of Ministers and participants from the academic world, national researchers and representatives from non-governmental organizations. It aims to address climate change in the light of the most recent scientific forecasts for the twenty-first century, and to accomplish the commitments under the UNFCCC.

Other developments during 2006–2009 were the completion of various studies that illustrated the existing potentials for reducing GHG emissions in the country.

The Research and Studies on Energy Programme at the University of Chile carried out a study titled 'Estimations of Energy Saving Potentials in the Various Sectors through Improved Energy Efficiency' (PRIEN, 2008). It developed aggregated and sector-specific indicators of energy efficiency. The aim was to evaluate the development of the use of energy resources in terms of its efficiency and to apply these indicators to the period between 1990 and 2006. In addition, an estimation of the theoretical improvements in energy efficiency was made in each sector and subsector.

Another study was commissioned by the company *Endesa Latinoamérica*, part of the Environmental Economics Management Programme at the University of Chile (PROGEA). The study was titled 'Energy Consumption and GHG Emissions in Chile 2007–2030 and Mitigation Options' (PROGEA, 2009). PROGEA forecast the country's GHG emissions up to 2030 and evaluated various policy instruments to reduce emissions.

In the study, the major GHG reduction measures were identified and evaluated for the following sectors; transport, commerce, public and residential, industry and mining and electric energy generation. In addition, the economic and regulatory instruments for promoting and implementing the proposed measures were identified. An estimation of the costs of implementation and reduction potentials were based on expert opinion and information from companies and regulatory bodies, using, *inter alia*, literature collected on the topic from national and international sources. The study was also supported by the scenario simulation tool LEAP, the Long range Energy Alternatives Planning System developed at the Stockholm Environment Institute.

Based on these developments and resulting studies, and with the opportunity provided by COP15 in Copenhagen in December 2009, the country, through the Ministry of Environment, made a national announcement of its goal to counter the trend of growing GHG emissions and achieve a 20% reduction compared to 'BAU' by 2020.

2.4 *Fourth phase (2010 to present)*

The fourth phase of climate change mitigation activities began with the communication of the pledge made in Copenhagen. The present period is characterized by: the formalization of the pledge by registering it in Appendix II of the Copenhagen Accord; an active process of identifying mitigation options in order to accomplish the announced goal, including NAMAs; and important national institutional developments for tackling climate change.

Importantly among these last developments was the enactment of the law that creates the Ministry of Environment (Law 20.417). This replaced the National Commission of the Environment which was the driving force of environmental policy between 1994 and 2009. It increased the level of political significance of environmental issues in the country and, in particular, established as a function of the ministry the proposal of 'policies and formulation of plans and programmes of action for climate change mitigation'. It stipulated that this should be done in collaboration with the different administrative bodies of the state on national, regional and local levels, in order to 'establish the effects of, and the necessary measures for, adaptation to and mitigation of, climate change'. The Climate Change Office in the Ministry of Environment is in charge of these functions. Broader oversight is provided by an Inter-Ministerial Climate Change Committee, consisting of the ministries of Environment, Agriculture, Housing, Foreign Affairs and the Presidential General Secretary. The committee was created in mid-2009 as a response to the need to coordinate Chile's position in international negotiations on Climate Change, particularly in light of the Conference of Parties held that year in Copenhagen, and has during the present period emerged as the highest ranking body dealing with climate change in the country.

Chile's commitment to act was formally communicated in a letter to the UNFCCC in August 2010 (Chile, 2010). Chile indicated that it will require a significant level of international support to achieve this commitment and that it would focus its actions on measures of energy efficiency, renewable energy, forest and land-use change.

Upon formalization of the commitment, research and studies to indentify plan and implement mitigation measure have been conducted, particularly aimed to the identification of NAMAs that allows for reaching the goal announced. These initiatives are being carried out by the authorities of sectors that are responsible for major GHG emissions and which, therefore, present great potentials for achieving important reductions. These are as follows: the Ministries of Energy, Transport and Agriculture. The Ministry of Environment is playing an active role in promoting these efforts and, correspondingly, the Foreign Affairs Ministry in searching for potential financial supports for the ideas of NAMAs that are emerging from these works.

The CONAMA conducted a study titled 'Analysis of Future GHG mitigation options for Chile in the Energy Sector' (POCH, 2010). Mitigation scenarios for a time-frame of 20 years (2010–2030) were constructed. They considered sectoral energy demands and supply options post-2000. The study defined technological mitigation options, options for generating energy and changing combustibles. The study was strengthened by the use of the scenario simulation tool LEAP. The study defined in detail a reference scenario for the energy sector during this period, taking electricity demand forecasts derived from economic growth into account, and also the current and forecasted capacity of the country's electricity grids.

However, the country has not conducted a formal discussion to achieve a common approach to NAMAs, nor to establish criteria for prioritization and the various aspects required for their implementation, including issues of measurement, reporting and verification (MRV) of their results. Under these circumstances, the activities that have been carried out so far have depended on the progress made in international negotiations on the matter in the arena of the UNFCCC.

3. Development of NAMAS in transport, agriculture and energy sectors

The current status of these initiatives is discussed in this section. Each initiative in the transport, agriculture and energy sectors is at a different phase of development in terms of formulating NAMAs. All of them have reached a first project phase of formulating clearly defined goals, setting estimates for reducing emissions and drawing up preliminary proposals for achieving them. Only the energy sector has drawn up proposals for other essential aspects concerning implementation. In the following sub-sections, when reporting percentage reductions in emissions, these are relative to a baseline specific to the action.

3.1 *Transport sector*

One of the NAMAs identified in the transport sector aims to promote energy efficiency in transport as an effective tool for achieving GHG emission reduction, ensuring that a sustainable balance between the system load and passengers is maintained. It comprises measures such as: aerodynamic improvements, with an estimate annual CO_2e reduction of 6% compared to baseline; training in efficient driving, with an estimated annual CO_2e reduction between 9% and 15%; good maintenance practices, with an estimated annual CO_2e reduction between 7% and 10%; and improved fleet (traffic) management, with an estimated annual CO_2e reduction between 5% and 15%. At the level of emission of CO_2e from the transportation sector in Chile at the year 2006, these figures would mean a reduction between 4600 and 6900 M tonnes per year (Salgado, 2011).

A second NAMA identified aims to generate incentives for an increase in zero and low emission vehicles in the vehicle fleet by continuing initiatives such exemptions from payment for circulation permits or implementation of new systems such as the 'feebates' system. The estimated annual reduction of these measures are between 2% and 3% per year, or between 340 and 510 M tonnes of CO_2e as compared with the emission from the transportation sector in Chile during the year 2006 (Salgado, 2011).

A third NAMA aims to promote a change in the mode of transport, from private to public transport and from motorized to non-motorized transport. This is to be achieved through measures as: promotion of non-motorized transport, with an estimated annual CO_2e reduction between 0.5% and 1%; design of bicycles bays in strategic places of services, with an estimated annual CO_2e reduction between 0.5% and 1%; and design and construction of underground parking places in the Metro system and intermodal stations, with an estimated annual CO_2e reduction between 1% and 2%. At the level of emission of CO_2e from the transportation sector in Chile at the year 2006, these figures would mean a reduction between 340 and 680 M tonnes per year (Salgado, 2011).

The fourth NAMA identified to date in the transport sector aims to implement measures to manage traffic in the cities, with the goal of optimizing road operation and at the same time mitigating CO_2e emissions. Main measures contemplated in this programme are: Analysis and development of centralized system of transit area, with an estimated annual CO_2e reduction between 5% and 7%; analysis and development of traffic calming area projects, with an estimated annual CO_2e reduction between 0.5% and 1%; and analysis and development of reversible, segregated and exclusive road lanes, with an estimated annual CO_2e reduction between 1% and 2%. At the level of emission of CO_2e from the transportation sector in Chile during the year 2006, these figures would mean a reduction between 1,100 and 1,700 M tonnes per year (Salgado, 2011).

3.2 *Agriculture sector*

The NAMAs indentified in the Agriculture Sector are based on information collected in two completed studies the year

2010: (i) an 'Analysis of future GHG mitigation options for Chile associated with programmes promoting the agriculture and forestry sector' (Pontificia Universidad Católica, 2010), conducted in collaboration with the Ministry of Environment and (ii) an estimation of the 'Potential climate change mitigation associated with the recuperation of native forest and forestry development law' (INFOR, 2010).

The aim of the first NAMA is to contribute to GHG emission mitigation in Chile through promoting sustainable management and rehabilitation of native forest in Chile. The proposal considers affecting 693.507 ha from now until 2050 with an increase in CO_2 capture of 234 M tonnes. At a total implementation cost of USD 716 million, the approximate mitigation cost is US$ 3.06 per tonne of CO_2 captured (Barrera, 2011)

The aim of the second NAMA is to contribute to mitigation of Chile's GHG emissions via promotion of afforestation in areas of degraded soils and with soils suitable for forestry. The proposal estimates an increase in capture of CO_2 between 8.4 and 29.2 M tonnes from now until 2020 and between 12 and 51 M tonnes since 2021 until 2030, and a cost per tonne of CO_2 captured between 12 and 43 US$/t$CO_2$e for the first period and between 11 and 35 US $/t$CO_2$e for the second (Barrera, 2011).

3.3 *Energy sector*

Five NAMAs of particular interest have been selected in the first round of selections of mitigation actions in the energy sector. Similar to the agriculture sector, these were developed based on information from previous studies on opportunities for mitigation actions in the country that were discussed in the previous section. Two of these NAMAs promote an increase in the integration of renewable energies in the country's energy grid. Three of them promote energy efficiency in the copper mining sector, the industry in general and the cement subsector (CCYD Consultants, 2011).

The first NAMA focuses on incentivizing efforts towards creating energy efficiency in one of the most important and productive sectors in the country and the main source of emissions in the mining sector, namely the copper industry. The final aim of this programme is to reduce CO_2 emissions in the sector by approximately 4.8% by 2020. Measures contemplates by this programme are: Replacement of traditional SAG mills crushers with high pressure crushers; use of Parabolic Solar Collectors and heat pumps to control temperature of electrolytes and other heat sources; and use of Bio fuel in the transport of mining material. With the three measures together, the estimated reduction is 6 million tonnes annually by 2020 (CCYD Consultants, 2011).

The second NAMA aims to reduce the intensity of energy use in industry and mining by 1% of what would exist without the programme by 2020. Measures contemplated are: Investment in efficient motors and replacement of obsolete motors (with more than two repairs); and installation of inverters for the functioning of pumps and ventilators. With these two measures, the expected reduction is 0, 36 million tonnes annually by 2020 (CCYD Consultants, 2011).

The third NAMA aims to increase renewable energies in the country's electricity grid. The sources considered in this programme are competitively priced and have not been developed due to economic or financial restrictions (mini-hydro, biomass and wind energy). The goal is to create an extra 850 MW of NCREs in Chile by 2020. This will result in a reduction of GHG emissions of 2 million tonnes by that year, calculated as if the programme did not exist (CCYD Consultants, 2011).

The aim of the fourth NAMA is to reduce emissions in the cement industry via energy efficiency and the use of alternative energies, reaching a reduction of 0.24 million tonnes of CO_2e from the 2020 forecasts. Measures contemplated are: replacement of fuels in cement ovens and increase of energy efficiency in systems of processing primary materials and in grinding/milling (CCYD Consultants, 2011).

The aim of the fifth NAMA is to incentivize the development of energy generation using geothermal energy through national policy. This to reach a potential capacity for 320 MW additional for the scenario based on 2020. Measures contemplated are: reduce obstacles entering into the national electric grids as far as possible; create temporary financial incentives to promote entry of certain technologies; and develop policies that consider externalities in decision-making in the electricity sector. With the three measures together, the reduction anticipated is approximately 3 million tonnes by 2020 (CCYD Consultants, 2011).

4. Systems to support NAMAs and realize co-benefits

As stated above, the NAMAs listed in Section 3 are all in different stages of development with the agricultural and transport sector NAMAs lagging behind the development of the energy sector NAMAs. None of the sectors, however, have made much progress in creating a dedicated institutional and governance framework that would facilitate rapid development or implementation of NAMAs. Thus, this section will outline the existing institutional and governance framework in the country that might conceivably be adapted and built upon to enhance the opportunities for NAMA development.

4.1 **Policies and regulations**

In the information available on the NAMAs listed in the previous section – with exception of those in the energy sector – there is no mention of the policies and regulations that support and sustain them. However, such policies exist but seem not to inform the early phase of development. In terms of the availability of the institutional capacity to design NAMAs, whilst the specific expertise does not appear to exist, Forestry (under Agriculture) has a long-standing technical expertise that may be drawn upon successfully and the Transport Sector may be able to do so with support from the Ministry of Environment.

This is different for NAMAs in the energy sector, as the topic has been given special attention in the preliminary documents and there exists the institutional capacity to design NAMAs within the free market system in the sector. Moreover, the identified NAMAs set major goals, in line with the goals set in governmental policies for achieving increases in the efficient use of energy in the country's electricity grid, as described in the previous section. Therefore, if these NAMAs are implemented, they will play an integral role in achieving government goals. Regardless, it will be necessary, as noted in the NAMA proposals, to develop additional regulatory frameworks in order to ensure that the goals of each are indeed achieved.

4.2 *Measurement reporting and verification and institutions required*

To date, only the NAMAs identified by the Ministry of Energy have made proposals for MRV on their progress. Drawing from the proposals, the characteristics of MRV systems for national consideration are as follows.

Since each NAMA is different and consists of a variety of actions, an MRV system needs to be tailored specifically to each. There is an abundance of literature and international experience on constructing national GHG emission inventories. These include inventories of countries, corporations and their production lines. There is also information on implementing systems for entering the Carbon Market and on good practices for using MRV systems. These should be used as guidelines for standards setting and designing NAMAs. It would be both useful to the process in terms of design and in finding international support for them.

It is in the government's interest to ensure that good practices are applied in monitoring and verifying the NAMAs. The baseline scenario and the plan for monitoring results during implementation need to be verified in terms of their soundness. In the case of verification of mitigation actions specifically, there are no prescribed guidelines, although general guidelines for domestic MRV are still under negotiation. Reporting on implementation of the NAMAs is important in the international system, to aid transparency. Evolution of the impacts that a NAMA has on patterns of GHG emissions in developing countries and annual progress in implementing actions must be tracked. The information must be sufficient for evaluating the effectiveness of a NAMA and in order to be able to assess whether it is wise to allocate resources to that action or not.

The private sector could, however, also initiate NAMAs that need international support. A national institution needs to be set up to process these initiatives. In addition to evaluating their appropriateness for national interests and circumstances, the baseline scenario, the monitoring process and the verification methods must be tested for soundness. A nationally recognized verification body must issue a technical report on these proposals. The national authority must keep a record of recognized verification bodies that can match the standards and capacities of internationally accredited systems, such as those in the Kyoto Protocol for verifying CDM results and the new standards in the ISO system (ISO 14.066).

The national authority should be also responsible for maintaining an online service, accessible to anybody, which publishes information on advancements and the results of NAMA activities. An additional advantage of this system is that it serves as a means for promoting proposals for NAMAs which seek international support.

4.3 *Financing*

Information on finance required is the weakest aspect of the NAMAs included in this article. This information is unavailable for the NAMAs in the Transport Sector. Information in the Agriculture and Energy Sectors is derived from previous studies on possible mitigation measures. The estimations made in each sector are subject to fluctuation in the long-term due to the fragility of the predicted scenarios. In addition, the scenarios are changing rapidly on both national and international levels.

Firstly, they are changing in the industrialized world due to new drive to introduce technologies in economies that will reduce dependence on fossils fuels, petrol on particular. Secondly, they are changing nationally due to the major energy policies currently in place in the country.

In addition, the administrative costs of the NAMAs have not been calculated. Nor is there information on what amounts of international aid will be sought. Despite these shortcomings, the NAMA design documents in the Energy Sector include a proposal for management of the financial resources required for implementation.

Additional research is clearly needed on the financing of NAMAs. There as been some preliminary work done on the design of specific financial instruments –revolving funds, concessionary finance, subsides and credits, for instance. The bulk of financial resources required is likely to be used to establish revolving funds for granting soft – loans to finance technological replacements that are in line with the NAMAs. For programmes promoting minor market entry of renewable energies, such as solar energy, part of the funds could be used to create subsidies to generate economic competitiveness of these technologies.

Thus, whilst there is an anticipated role for public–private partnerships, none have yet materialized.

4.4 *Socio-economic co-benefits*

The socio-economic benefits of NAMAs are often considered the ‘co-benefits’ of mitigation. They could also be called the developmental benefits, and are important in developing countries like Chile.

In the transport sector, measures to improve efficiency in management of transport could have important co-benefits for public health by contributing to a reduction of local air pollution, which is a growing problem in the main cities.

In the energy sector, the methodological approximation used for identifying and prioritizing the NAMAs considers in detail the socio-economic impacts that they may have. Therefore, in the design document, a proposal is made for measuring and reporting a group of indicators which reflect these impacts. A proposal is also made for the inclusion of these in the MRV system that the country is to establish.

The information on the NAMAs identified in the Ministries of Agriculture and Transport do not make mention of the socio-economic co-benefits of NAMAs. However, the implementation of the rigorous plans for afforestation proposed in the NAMAs in the agriculture sector would have significant positive impact on employment in the sector.

5. Conclusions

The purpose of this paper, as indicated in the introduction, is to describe the progressive involvement of the country in mitigation actions being divided into four periods Chile’s evolving efforts to meet its obligations under the Convention and make use of opportunities provided. The assessment does not try to evaluate past behaviour as appropriate or not, but rather assesses whether the present responses by Chile are sufficient in relation to the signals stemming from the Convention.

From the information presented in this article, it is evident that the development of NAMAs is emergent. The identified NAMAs are still in a stage of conceptualization and of making initial estimations on impacts and costs. The article has presented specific NAMAs in the NAMAs in the transport, agriculture and energy sectors.

In a few cases, attempts have been made to develop the NAMAs beyond the initial stage and some institutional development has taken place. Early experiences with institutional regulatory requirements on management and financing and on aspects of MRV have been made. This information can form a base for a more structured approach to NAMAs in Chile.

This article has found that the approach to NAMAs has been based on developments made in the context of the negotiation framework on a new international regime for addressing climate change. The socio-economic co-benefits (or developmental benefits) of NAMAs are likely to be important in further developments.

There is, therefore, an urgent need to reach some consensus around NAMAs in the country so that they can be finalized and integrated into national strategies. National strategies need to clearly affirm the motivation for the NAMAs both nationally and internationally, which is to support the current or future mitigation actions and to the commitment to reducing GHG emissions. There are multiple benefits to this approach: development of communal services; introduction of a framework for aspects of NAMA MRV; clarity on the financial support necessary and of the role played by the country in each case. Most importantly, it expresses political will by the government to move forward with NAMAs. Without strong Governmental commitment to implementing the NAMAs, it will not be possible to overcome the obstacles that threaten innovative programmes.

Note

1. An extensive review of the same can be found in the Country’s recently published Second National Communication to the UNFCCC (Ministry of Environment, 2011).

References

Barrera, D. (2011). NAMAs in the Agriculture sector. (J. E. Sanhueza Flores, Interviewer).

CCYD Consultants. (2011). *Identification, assessment, reporting and verification system for national mitigation actions for climate change in the energy sector*. Santiago: Author.

Chile. (2010). *Letter to the UNFCCC executive secretary 23 August, Minstry of Foreign Affairs*. Retrieved from http://unfccc.int/files/meetings/cop_15/copenhagen_accord/application/pdf/chilecphaccord_app2.pdf

Global Environmental Facility. (1992). *GEF*. Retrieved from http://gefonline.org/projectDetailsSQL.cfm?projID=372

Global Environmental Facility. (1996). *GEF*. Retrieved from http://gefonline.org/projectDetailsSQL.cfm?projID=270

Global Environmantal Facility. (1999). *GEF*. Retrieved from http://gefonline.org/projectDetailsSQL.cfm?projID=843

Global Environmental Facility. (2001). *GEF*. Retrieved from http://gefonline.org/projectDetailsSQL.cfm?projID=1321

Global Environmental Facility. (2003). *GEF*. Retrieved from http://gefonline.org/projectDetailsSQL.cfm?projID=1349

Global Environmental Facility. (2009). *GEF*. Retrieved from http://gefonline.org/projectDetailsSQL.cfm?projID=4136

Global Environmental Facility. (2010). *GEF*. Retrieved from http://gefonline.org/projectDetailsSQL.cfm?projID=4176

INFOR. (2010). *Potential climate change mitigation associated with the recuperation of native forest and forestry development law*. Santiago: Author.

Ministry of Environment. (2008). *Ministry of environment*. Retrieved from http://www.mma.gob.cl/1304/articles-49744_Plan_02.pdf

Ministry of Environment. (2011). *Ministry of environment*. Retrieved from http://www.mma.gob.cl/1304/w3-article-46280.html

POCH. (2010). *Analysis of future GHG mitigation options for Chile in the energy sector*. Retrieved from the Chilean

National System of Environmental Information website: http://www.sinia.cl/1292/articles-50188_recurso_6.pdf

Pontificia Universidad Católica. (2010). *Analysis of future GHG mitigation options for Chile associated with programmes promoting the agriculture and forestry sector*. Santiago: Author.

PRIEN. (2008). *Estimations of energy saving potentials in the various sectors through improved energy efficiency*. Retrieved from PRIEN website: http://www.prien.cl/documentos/caract_pot_ener_sectores.pdf

PROGEA. (2009). *Energy consumption and GHG emissions in Chile 2007–2030 and mitigation options*. Retrieved from Energycommunity website: http://www.energycommunity.org/documents/Aplicacion%20de%20LEAP%20en%20Chile,%202010.pdf

Salgado, P. (2011). NAMAs in the transport sector. (F. A. Ladron de Guevara A., Interviewer).

Sanhueza Flores, J. E. (2003). *National strategy study for the CDM in Chile*. Retrieved from CCYD website: http://www.ccyd.cl/publicaciones/informes%20de%20proyectos/nss.pdf

UNFCCC. (1992). *UNFCCC*. Retrieved from www.unfccc.int

UNFCCC. (1997). *UNFCCC – CDM*. Retrieved from http://cdm.unfccc.int/

UNFCCC. (2007). *UNFCCC*. Retrieved from www.unfccc.int

A case study of Peru's efficient lighting nationally appropriate mitigation action

Pia Zevallos[a], Talia Postigo Takahashi[a], Maria Paz Cigaran[a] and Kim Coetzee[b]

[a]*Libélula, Calle Alfredo León 211, Miraflores, Lima 18 Perú;* [b]*Energy Research Centre, University of Cape Town, Cape Town, South Africa*

This article seeks to begin to understand how mitigation actions (MAs) are approached and conceptualized in Peru by analysing this developing country's most advanced nationally appropriate mitigation action (NAMA) – the efficient lighting NAMA – in order to gain insight into how NAMAs are conceptualized as a vehicle for advancing MA in the country. Based on an earlier in-depth research report, this case study finds that the efficient lighting NAMA, despite being the most advanced in Peru, is still in the 'readiness' phase and as such more work and support is needed to move this NAMA and others to design and implementation phase. In the discussion, suggestions are made regarding how these implementation barriers might be overcome through increased coordination between key stakeholders in and across sectors, support for innovative approaches to financing and more clarity on monitoring, reporting and verifying issues.

Peru's perception of climate change is evolving from that of a highly vulnerable developing country prioritizing adaptation and its support, to that of a growing economy beginning to conceptualize low emissions development as an opportunity for increasing sustainability and tackling major structural problems such as poverty, low technology development, environmental degradation and conflict. However, at the time of writing, there are still no Low Emission Development Strategies (LEDSs), national or sectoral umbrella strategies/plans to provide guidance for sector-level mitigation actions (MAs). Given this context, this article presents and discusses the main findings of a study focusing on the state of climate change MAs (Takahashi, Zevallos, & Cigaran, 2011) in Peru by analysing Peru's most advanced nationally appropriate mitigation action (NAMA) – the efficient lighting NAMA – in order to gain insight into how NAMAs are conceptualized as a vehicle for advancing MA in the country.

The article proceeds by first providing a background of Peru's evolving MA strategy before analysing the efficient lighting NAMA as an in-depth case study. The case study approach highlights issues that may either assist or hamper the successful implementation of a NAMA. These issues are then discussed with a view to suggesting ways of improving future NAMAs.

1. Background: Peru's evolving mitigation action strategy

As a party to the United Nations Framework Convention for Climate Change (UNFCCC) and signatory of the Kyoto Protocol, Peru's national mitigation position is that 'mitigation has potential economic and social benefits for Peru, compared with the costs of reducing emissions' (Ministerio del Ambiente 2010b, p. 29). The Peruvian economy has produced a continuous annual average economic growth rate of 5.7% between 2001–2010 (Peruvian Central Bank, 2011) and this growth is linked to an increase in greenhouse gas (GHG) emissions and the intensification of socio-environmental conflicts; with these are a growing awareness of the need for sustainable development (Ministerio del Ambiente, 2010a). This awareness is evidenced by the recent inclusion of climate change as a competitiveness issue in the 2011–2013 Multi-annual Macroeconomic Framework (Ministerio de Economia y Finanzas, 2010) which is produced by the Ministry of Economy and Finance as the primary national planning instrument for public investment.

The Ministry of Environment's (MINAM) technical team has spearheaded domestic work on mitigation. The Action Plan for Climate Change Adaptation and Mitigation (Ministerio del Ambiente, 2010a), and the National Guidelines for Climate Change Mitigation propose a national mitigation effort composed of National Mitigation Programs (PRONAMIs). PRONAMIs are to be developed in the following sectors: forestry and land use, waste, energy, agriculture, transportation, industry and housing (Ministerio del Ambiente, 2011a,b). A MINAM commissioned study identified energy efficiency as the best sub-sector in which to pilot a NAMA. The study considered a range of factors including, but not limited to, potential emission reductions;

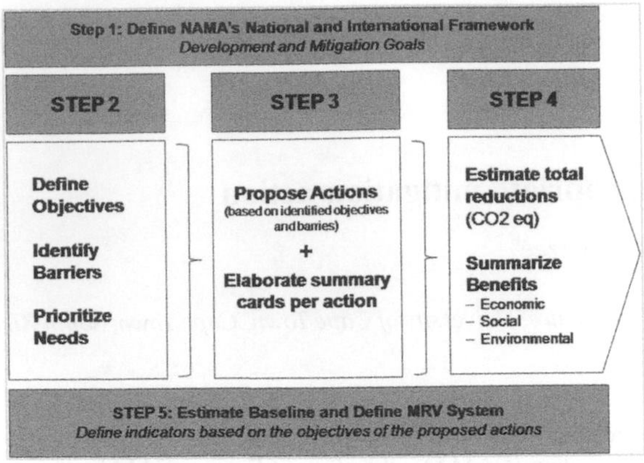

Figure 1. Methodological approach to formulating NAMAS. Source: Eslava et al., 2012.

the effectiveness and efficiency of implementation; economic and technical feasibility; alignment with policy priorities and socio-cultural conditions. As a result, the development of an efficient lighting NAMA was approved in the National Plan for Environmental Action (2010–2021) (Ministerio del Ambiente, 2011a,b) and a preliminary design exercise (Eslava et al., 2012) was undertaken. The following section analyses this efficient lighting NAMA in depth.

2. The efficient lighting NAMA case study

2.1. *Peru's approach to NAMAs*

In order to design NAMAs, a five-step methodology was proposed to catalyse the implementation of mitigation actions; this is outlined in Figure 1. The efficient lighting NAMA is still currently in concept stage, but this design methodology has so far been used by other stakeholders attempting to design NAMAs.

2.2. *Overview of Peru's proposed NAMA*

MINAM approached the Energy Efficiency Directorate of the Ministry of Energy and Mines to develop the NAMA proposal as it aligns with the Directorate's national priorities. In addition, a necessary legal framework already exists to support and promote the implementation of energy efficiency programmes and Peru already has experience of efficient lighting programmes in the residential sector. As part of the NAMA development, several coordination meetings and workshops with key stakeholders were conducted.

The proposed efficient lighting NAMA was made to fit under a conceptual framework previously developed, and consists of three MAs and three support actions; these are graphically represented in Figure 2. The MAs (columns) are intended to reduce energy consumption and facilitate the use of more efficient lighting technologies in the residential, industrial and public service sectors. The supporting actions (rows) were proposed to address the main barriers to the implementation and continuity of MAs identified in the exercise. Therefore, they should be oriented to promote private sector involvement, to increase the knowledge of the general public about efficient types of lighting and energy use good practice, to remove barriers to funding energy efficiency investment projects and to develop the necessary regulations for the implementation of efficient lighting technologies and a monitoring, reporting and verifying (MRV) system (rows).

The far right column of Figure 2 shows the possible cumulated benefits if all mitigation and support actions are implemented. The main goal of the efficient lighting NAMA is to reduce 4409.789 tons of CO_2eq in 10 years through reducing demand for power by 2779 mega watts and energy consumption by 80,109.2 Gigawatt hours in 10 years. Reduced power consumption is associated with lower CO_2 emissions. Other potential benefits include the increase of general public awareness regarding efficient lighting, the promotion of an energy saving and environmental care culture and poverty reduction. Also, efficient lighting is a good starting point to open up a discussion about CO_2 mitigation as it relates this more unfamiliar concept to a more familiar one of increased efficiency (Eslava et al., 2012).

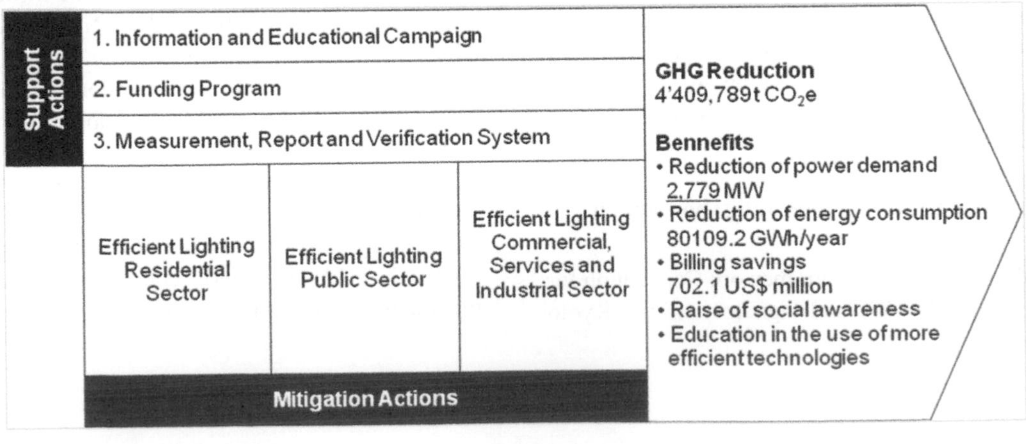

Figure 2. Efficient lighting NAMA scheme.
Source: Eslava et al., 2012.

2.3. *Analysis of the proposed efficient lighting NAMA*

2.3.1. *Study method*

This article presents the main findings of a study of MAs in Peru, undertaken for the Mitigation Action Plans and Scenarios (MAPS) programme in 2011 (Takahashi et al., 2011). The full MAPS study was based on a survey and analysis of available literature from the government, as well as of secondary literature by academic researchers and private sector consultants contracted by the government. Semi-structured interviews were conducted during workshops where the efficient lighting NAMA was presented to a cross-section of stakeholders.

The analysis of the proposed efficient lighting NAMA identified several areas of potential concern for its successful implementation. These include the existing planning, policy and regulatory context, institutional and technical capacity to design NAMAs, and the institutionalization of funding; these will be discussed in more depth below. Whilst the case study method necessarily limits the generalization of lessons learnt to other NAMAs in other sectors, the in-depth analysis is considered a useful basis for analysis of further MAs when they are developed in Peru.

2.3.2. *Planning, policy and regulatory context for energy efficiency*

The efficient lighting NAMA design exercise is the first of its kind in the country to be developed in a participatory process through open consultation. This is quite unlike the development of other long-term plans (including plans for the expansion of the energy system; energy efficiency and consumption) produced by the Ministry of Energy and Mines. Figure 3 captures the main stakeholders identified in the efficient lighting NAMA design exercise and highlights the considerable horizontal and vertical policy and policy integration that will be required to facilitate the PRONAMI.

The formulation of national and sector policies typically respond to short-term, or at best, medium-term interests, as the long-term view usually exceeds the period of a presidential mandate. The constant turnover that occurs at managerial and technical levels of the administration severely hampers on-going capacity building efforts and undermines the country's technical capacity to develop programmes and projects as key personnel are deployed elsewhere.

Many policies and regulations have been created in the preceding 10 years to promote and support energy efficiency actions and programmes. None, however, were specifically designed to provide funding or help the projects to obtain funding from private or international sources. In addition, none of the policies and regulations were geared towards increasing the national technical capacities to implement the projects. Additionally, not all decision-makers are well informed about the interaction of human activities, GHG emissions and the impacts of climate change. For example, many of the energy sector initiatives are carried out without consideration of their potential to increase, reduce or even avoid emissions.

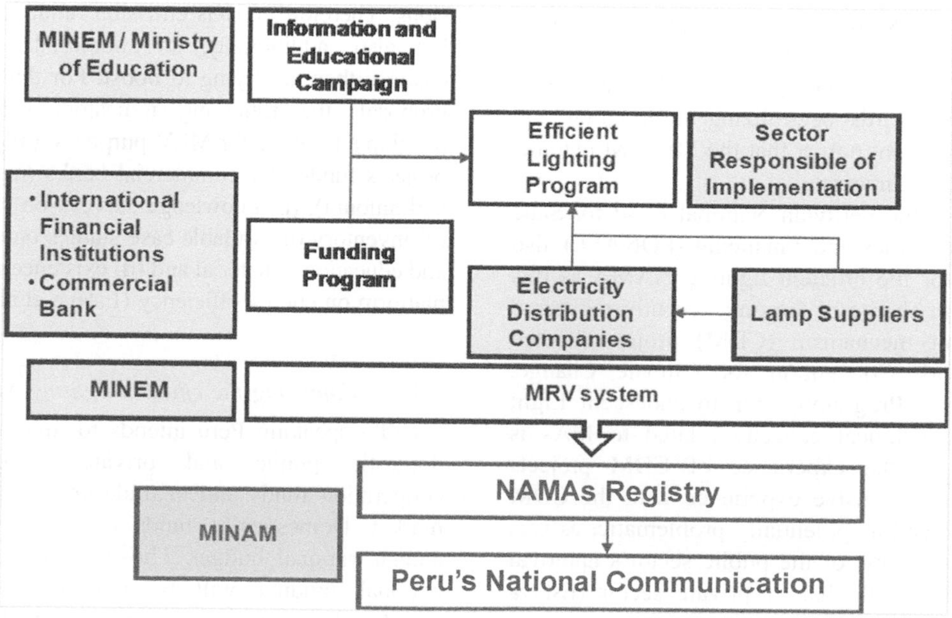

Figure 3. Stakeholders' interaction scheme for efficient lighting NAMA.

2.3.3. *Institutional & technical capacity to design and implement the efficient lighting NAMA*

The main challenge Peru faces when designing MAs is the lack of reliable national information systems and capacity to create projection models (Euroconsult Mott Macdonald, 2010). The national information systems are not centralized, the information is outdated and there is a lack of harmonized methodologies for evaluating the economic growth and emissions reduction potential of different NAMAs, in order to prioritize their development (MAPS Peru, 2010). A low 'coordination culture' and the difficulties of transferring of authority between the relevant institutional stakeholders were identified as having the potential to undermine the successful implementation of the efficient lighting NAMA (Euroconsult Mott Macdonald, 2010).

As a relatively new authority, the MINAM (created in 2008) has acquired some responsibilities and competences previously exercised by other Ministries and public entities. This is the case of the new Organisation for Environmental Evaluation and Control (OEFA) attached to MINAM, which became part of the National System for Fiscal and Environmental Regulation (Law N°29325) in 2009. OEFA's main objective is to enact environmental legislation as well as to supervise and assure the effective evaluation, supervision, fiscal control and sanction in environmental matters.

In a future context of mandatory mitigation, OEFA would also assume important MRV functions (discussed more below) as thus far Peru also lacks a national system for monitoring environmental programmes and projects. The transfer of the environmental inspection responsibility from sector agencies like OSINFOR (Forests) and OSINERGMIN (energy) to the newly created OEFA has generated some difficulties in relation to the national environmental control system and has temporarily affected its monitoring, control and enforcement capacities due mainly to a loss of experienced personnel. However, it is a necessary step towards centralizing information and building the enforcement power that the MINAM requires to fulfil its monitoring mandate.

The capacity of the Peruvian National Fund for State Entrepreneurial Activities and Financing (FONAFE) also has implications for the efficient lighting NAMA as this agency has been in charge of the only recently registered clean development mechanism (CDM) project (United Nations Framework Convention for Climate Change, 2010), the National Programme for Incandescent Light Replacement. As technical capacity related to MAs is thus far limited to the experience of CDM projects FONAFE's lack of extensive experience with the CDM process was perceived as potentially problematic, as was the greater length of time of the public sector's approval processes compared with that of private sector (Eslava et al., 2012).

In addition to coordination requirements mentioned in 3.2 above, and despite the creation of the Directorate General for Energy Efficiency within the Ministry of Energy and Mines, there was still a perceived lack of knowledge of new technological alternatives for energy efficiency.

2.3.4. *MRV systems for NAMAs in the energy sector*

All information related to activities in the energy sector is centralized within the Ministry of Energy and Mines. The General Office of Planning in the Ministry operates a system that ensures consistency, understanding, accuracy and transparency. The only requirement to satisfy the Peru's GHG inventory system is to improve comparability, which would be achieved by incorporating GHG data to the existing National Energy Information System (Ministerio del Ambiente, 2010a).

The existing National Energy Information System does not represent a complete picture of the mitigation activities in Peru. Most of the MAs are not measured by it or properly accounted for, and are therefore also excluded from the country's National Communication to UNFCCC. Only registered CDM projects have an appropriate monitoring and reporting system that allows progress and emission reduction to be tracked. However, current efforts are being developed by MINAM to implement a National Information System for GHG accounting, called SINENGEI.

In addition to the potential benefits of a National Information System and to further address the gaps in reporting, the efficient lighting NAMA proposes a different baseline and indicators for the NAMA MRV system. MINAM and MINEM have concluded that NAMAs should promote transformational changes and, at the same time, reduce emissions. Therefore, GHGs emission reductions be measured, but also, additionally, so should the transformational changes they are trying to boost. For this specific NAMA proposal, the following indicators were proposed to develop a baseline for MRV purposes: (a) energy efficiency projects funded by commercial banks (number of projects and amount); (b) knowledge surveys on energy efficiency; (c) inventory of available case studies on energy efficiency and educational material and (d) existence of an information platform on energy efficiency (Eslava et al., 2012).

2.3.5. *Financing the efficient lighting NAMA*

Broadly speaking Peru intends to finance its MAs using domestic public and private funds, international cooperation funds and to trade emissions on the carbon market. Domestically, funds will be allocated from the annual national budget. Those NAMAs that seek international funding will be placed in the UNFCCC's NAMA Registry (in accordance with the decisions adopted during COP16 in Cancun). Peru would seek funding through current carbon market mechanisms as well as any new mechanisms established through future UNFCCC negotiations (Takahashi et al., 2011).

For the preparatory phase of this NAMA, Peru requires technical assistance and international financial aid in order to complete the design and also follow its implementation. In particular, all the support actions (Figure 2) of the efficient lighting NAMA require both technical assistance and international funding to be effectively implemented.

The NAMA methodology design exercise found that it was difficult to involve commercial banking institutions in the financing of energy efficiency projects as they do not consider the future savings generated by this kind of project as sufficient guarantee to assume risk in the present by providing funding at a reasonable interest rate. Therefore, international technical assistance in building capacity for the commercial banking is also needed. The priority areas for international cooperation identified in the study are: the establishment of the baseline and the MRV system, support for the implementation of GHG Report (the network of national GHG inventory), including updating the inventory of 2000 and the support in the design and implementation support to the financing programme to catalyse public and private investments (Eslava et al., 2012).

The Peruvian approach to climate change mitigation is to increasingly mainstream climate change MAs in its development planning strategies. To date however, there remains no overarching national or sectoral level strategy to guide or coordinate these ad hoc MAs. Nevertheless, Peru has shown noticeable progress in the past 10 years, most recently by developing its own NAMA methodological approach and identifying how MAs could complement current national development plans. The paper argues climate change mitigation is still not a priority for sectorial Ministries despite MINAM's efforts to foster collaboration. Therefore, NAMA development in the energy sector should focus on building on existing priorities: increased efficiency in industry and increasing environmental law enforcement to avoid social conflict. It should be led by experts of these fields, who would quantify mitigation potentials in collaboration with mitigation experts.

This case study highlighted that even in a sub-sector prioritized because it was considered possible to initiate immediate and cost-effective MAs, there remain significant cultural and financial barriers to implementation. The very weak culture of energy saving and the complexities of the issue combined with a low 'coordination culture' between stakeholders all provided barriers. The NAMA exercise identified the need to involve a wide variety of stakeholders early on in the NAMA design phase in order to develop a successful and feasible programme. This task requires a high level of coordination in a country where the national coordination culture is very low. This strategy of inclusion and coordination is necessary from the beginning of the design phase, however, in order to avoid (or lessen) the risk of rejection by a relevant stakeholder during the implementation phase. In this particular case, the early involvement of the Ministry of Energy and Mines was crucial to identifying key barriers and opportunities.

In addition, a variety of financial barriers exist: the most significant are the low-energy prices and the lack of information available to potential investors which increases the perceived risks of these of large-scale investments. Another financial barrier is that most funding schemes provide credit against the projection of higher incomes but not against the projection of future savings. Tariffs set by the regulator are also a topic to explore, since there could be room to promote efficiency by applying differentiated economic incentives.

The proposed efficient lightning NAMA has a clear policy context; it is based on an existing Energy Efficiency Plan (Ministerio de Energía y Minas, 2009) updated during 2011. The former plan was not ready for execution as it had not identified funding sources, allocated funds, defined MRV systems or private sector involvement. It is, however, the only sector policy that explicitly addresses emission reductions by roughly calculating the possible emission reduction benefits. The case study reveals that climate change mitigation is still not a priority for sectoral Ministries despite MINAM's efforts to foster collaboration. Therefore, NAMA development in the energy sector should focus on building on existing priorities: increased efficiency in industry and increasing environmental law enforcement to avoid social conflict. It should be led by experts of these fields, who would quantify mitigation potentials in collaboration with mitigation experts so as to bring the NAMAs developed into line with Peru's developmental goals.

The efficient lightning NAMA case study also illustrates the need for consideration of the concept of 'readiness': preparatory activities that create an enabling environment where NAMAs can be subject to MRV and to which low-carbon investment can be directed. Whether these preparatory activities take place in the context of the development of a specific NAMA or in a more systematic approach, for example, as part of the development of LEDS, would be a key area requiring further research.

The proposed approach to MRV in Peru goes far beyond measuring and reporting emission reductions. It includes the monitoring and measurement of indicators of transformational changes and gives special attention to measuring benefits of mitigation. It also includes a registry of financial resources invested and involved stakeholders. Although lighting efficiency programmes in the residential sector can catalyse transformation, reductions are very difficult to measure; confidentiality issues and the highly informal nature of the industrial sector serve to hamper effective reporting. Therefore, the analysis suggests that a culture of disclosure needs to be developed within the industry sector. Furthermore, for the purposes of MRV, there is an opportunity to apply lessons learned from CDM, for instance the development of baseline estimation methodologies, and the design of institutional arrangements.

It is clear that for Peru, MAs should always be nationally appropriate. In other words, NAMAs should be transformational: designed not only to reduce emissions, but also to promote significant changes in behavioural patterns, levels of investment, cultural assimilation, availability of information and technical capacity. One of the key aspects emerging from the analysis of the efficient lighting NAMA proposal is that NAMA development should be considered as a way of defining and funding actions that identify and overcome barriers to effective implementation of MAs. Although NAMA development in Peru is at a very initial stage, two main avenues to advance MA are identified: one, work under the Clean Development Mechanism; two, policies and measures that predominantly have mitigation as the co-benefit, (instead of the main driver). The efficient lighting NAMA for instance – designed with support activities to overcome barriers, to gain international funding and to be 'MRVable' – has the potential to produce more co-benefits than emissions reductions.

Finally, it is important to note that this is the first exercise of formulating a NAMA for Peru, and that part of the purpose of the original study on which this paper is based was to identify the steps to complete its design. Since it is a NAMA, it has the potential to be implemented in the short term and pave the way for replication of the successfully implemented aspects.

References

Congreso de la Republica. *Ley N°29325 Ley del Sistema Nacional de Evaluacion y Fiscalizacion Ambiental*. Lima, March 2009. Retrieved August, 2011, from http://www.congreso.gob.pe/ntley/Imagenes/Leyes/29325.pdf

Eslava, J., & Millán, R. (2012). *Prioritization and formulation of a mitigation action*. PPT presentations from José Eslava (MINEM) and Rafael Millán (MINAM) at MAIN Video Conference Session May 24, 2011.

Euroconsult Mott Macdonald. (2010). *Developing countries, monitoring and reporting on greenhouse gas emissions, policies and measures – Country Report Peru*. Retrieved August, 2011, from http://ec.europa.eu/clima/policies/g-gas/monitoring/docs/country_report_peru_2010_en.pdf

MAPS Peru. (2010). *Estado del Arte*. Documento Preliminar aprobado por el Comité Directivo de Cambio Climático del Ministerio del Ambiente, Lima, Peru.

Ministerio del Ambiente. (2010a). *Plan de Acción de Adaptación y Mitigación frente al Cambio Climático*. Retrieved August 3, 2011, from Sistema Nacional de Información Ambiental: http://sinia.minam.gob.pe/

Ministerio del Ambiente. (2010b). *Segunda Comunicación Nacional del Perú a la Convención Marco de las Naciones Unidas sobre Cambio Climático 2010*. Lima: Fondo Editorial del MINAM.

Ministerio del Ambiente. (2011a). *Plan Nacional de Acción Ambiental*. Retrieved August, 2011, from http://www.minam.gob.pe/index.php?option=com_content&view=article&id=871

Ministerio del Ambiente. (2011b). *Propuesta de Lineamientos Nacionales de Mitigación. Documento Preliminar*. Retrieved January 25, 2011, from http://cdam.minam.gob.pe:8080/dspace-consultorias/bitstream/123456789/162/2/CD000106-P.pdf

Ministerio de Economia y Finanzas. (2010). *Marco Macroeconomico Multianual 2011–2013*. Lima, Peru.

Ministerio de Energía y Minas. (2009). *Plan Referencial de Uso Eficiente de la Energía 2009–2018*. Retrieved February 25, 2011, from www.minem.gob.pe

Takahashi, T.P., Zevallos, P., & Cigaran, M.P. (2011). *Country Study on Mitigation Actions in Peru. Research Report for MAPS (Mitigation Action Plans and Scenarios). Lima, Peru, Libelula*. Retrieved from http://www.mapsprogramme.org/wp-content/uploads/Final_Country-Study_MA_Peru_111124.pdf

United Nations Framework Convention for Climate Change. (2010). *CDM: Project Cycle Search, Peru*. Retrieved October 2010, from http://cdm.unfccc.int/Projects/projsearch.html

A case study of South African mitigation actions

Emily Tyler[a], Anya Sofie Boyd[b], Kim Coetzee[b] and Harald Winkler[b]

[a]*Independent Researcher, Rosebank, Cape Town, South Africa;* [b]*Energy Research Centre, University of Cape Town, Cape Town, South Africa*

This article seeks to understand how mitigation actions (MAs) are approached and conceptualised in South Africa, and then to capture the particular sets of issues and characteristics relating to these actions. As such it considers three main areas of enquiry from a bottom-up methodological perspective: first, what is the South African approach to individual MAs, second, what are the barriers and challenges to their implementation, and third, what by way of domestic measures and international support could assist in overcoming these challenges. Four examples of potential South African MAs are described and then analysed: the Bus Rapid Transport in Cape Town, the South African Renewables Initiative, the carbon tax and the National Sustainable Settlements Facility. We find from considering these examples that there are significant challenges to defining an MA. We also find that, generally, South Africa is good at identifying, analysing and designing activities to mitigate emissions, but lacks in effective implementation. Two main areas of implementation risk are suggested, namely, counteracting vested interests and the availability of finance. Suggestions are made regarding how these implementation challenges might be overcome with appropriate support at the domestic and international levels.

1. Introduction

As part of this Special Issue exploring the concept of mitigation actions (MAs) in developing countries, this South African case study seeks to understand how individual MAs are approached and conceptualised in the country and then to capture the particular sets of issues and characteristics relating to these actions.

Establishing a precise definition of an MA for the purposes of the analysis proved distractingly complex. While some of the issues around definition arising from consideration of individual MA examples are briefly discussed in the paper, the matter is left unresolved. Therefore, the term MA is used here as a working definition, loosely describing initiatives which have a strong impact on reducing emissions.

The paper has three main areas of enquiry:

1. What is the South African approach to MAs?
2. What are some of the barriers and challenges to the implementation of MAs in South Africa?
3. What by way of domestic measures and international support could assist in overcoming these challenges?

The paper draws its findings from both a consideration of the country's mitigation context and policy approach to individual MAs, and from an analysis of four MA examples. The paper proceeds as follows. The methods used are described in Section 2. A general overview of how MA is being undertaken in South Africa provides a context in Section 3. In Section 4, four non-representative examples of potential MAs are described: the Bus Rapid Transport (BRT) in Cape Town, the South African Renewables Initiative (SARi), the carbon tax and the National Sustainable Settlements Facility (NSSF). In Section 5, the four examples are analysed to see what can be discovered about mitigation activity in the country. Section 6 discusses the findings and concludes.

2. Methodology

The research in this article builds on a country study conducted under the Mitigation Action Plans and Scenarios (MAPS) programme, and benefitted from interaction with similar studies undertaken in other MAPS countries, reported in this Special Issue. The studies did not use a particular theoretical framework, but rather a common set of research questions.

The analysis proceeded by establishing the South African mitigation context, and choosing and describing four examples of South African MAs. The examples were randomly chosen from the suite of prominent MAs in the country, with an emphasis on scale and emission reduction impact, but also on the inclusion of a diversity of actors, sectors and type of mitigation initiative. The research was conducted by the authors between May and July 2011, and was based on the authors' in-field knowledge and experience, and comprised desktop literature reviews and two semi-structured interviews with key people associated with the chosen examples.

A set of criteria was then developed to consistently analyse the four examples against the study's three areas of enquiry. The criteria were developed by firstly drawing from previous MAPS work[1] on NAMAs as well as existing sources from international think-tanks like Ecofys, before posing ourselves a range of questions to broaden our understanding of how MAs are, or might be, defined, supported and potentially implemented in South Africa. The MA examples were analysed against the criteria, and conclusions were drawn.

The bottom-up method of interrogation of examples was primarily adopted to deepen our understanding of mitigation activity in the country. Although the authors acknowledge the limitations of generalising from such a small and non-representative sample, this method is, nevertheless, held to serve the overarching purpose of opening up the discourse on MA from a developing-country perspective.

3. South African mitigation activity: a brief overview

South Africa is the largest single economy in sub-Saharan Africa, with the highest CO_2 emissions on the African continent. Even though South Africa accounts for only 1.19% of the total world CO_2 emissions, the high carbon intensity of the economy means that it ranks 44th out of 185 countries infor the per capita emissions of CO_2 (CAIT, 2011).

Internationally, South Africa committed, under the Copenhagen Accord, to 'take nationally appropriate mitigation action to enable a 34% deviation below the "Business As Usual" emissions growth trajectory by 2020 and a 42% deviation below the "Business As Usual" emissions growth trajectory by 2025 .... The extent to which this action will be implemented depends on the provision of financial resources, the transfer of technology and capacity-building support by developed countries ...' (RSA, 2010). This ambition has become known informally within the country as the 'Peak, Plateau and Decline' (PPD) trajectory. As the custodian of the Durban Platform for Enhanced Action negotiated under the United Nations Framework Convention on Climate Change in Durban in 2011, the country is increasingly expected to align its domestic MAs with an ambitious international position on mitigation.

South Africa's seminal endeavour to identify the suite of mitigation opportunities available to it occurred through the 2006–2008 Long Term Mitigation Scenario (LTMS) process conducted for the Department of Environmental Affairs and Tourism. The LTMS used the Pacala–Socolow wedge method (Pacala and Socolow, 2004), describing and quantifying 10 large, 13 medium and 9 small technological and economic policy instrument wedges to reduce the country's emissions between 2010 and 2050. Many actions will be required to effect these and other emission reductions in South Africa. These will necessarily be very diverse, and potentially include policies, strategies, targets, voluntary agreements, regulation, standards, economic instruments, financial mechanisms, subsidies, programmes, projects, pilots, market initiatives, capacity development, information generation, innovation, institution-building, Centres of Excellence, partnerships, skills development and more. The question of whether all of these initiatives (or indeed the LTMS wedges themselves) should be referred to as MAs has not yet been formally posed in the country.

Policy discourse in South Africa so far has neither yielded a definition of an MA nor a detailed support strategy for these initiatives. The National Climate Change Response White Paper (RSA, 2011a) suggests that MA will result in the country adhering to the PPD trajectory, that different types of mitigation approaches, policies, measures and actions will be used, that actions will be identified at a sector and a sub-sector level, and that companies and sectors will be required to submit mitigation plans. It does not, however, define MA, nor does it formally establish this concept. A carbon budget approach is proposed to frame South African mitigation activity going forward, and the combination of actions incentivised will be identified based on incremental and direct costs, and broader socio-economic development impact and international competitiveness (RSA, 2011b). A suite of near-term priority flagship programmes is identified by the White Paper, including a number in all the major emitting sectors. It also indicates a framework for institutional co-ordination of mitigation opportunities to include parliament, the Inter-Ministerial Committee on Climate Change, the Forum of South African Directors General and the Intergovernmental Committee on Climate Change, with MINMEC and SALGA providing guidance on the provincial and local level.

Climate mitigation is also addressed in the climate change chapter of the National Planning Commission's First Development Plan (launched in August 2012). However, there are problems of consistency between the framing of mitigation targets and approach in this chapter and the Climate Change Response White Paper. The chapter supports the need for MA, emphasising the context of economic development and the realisation of the policy objectives of poverty alleviation and employment creation. It is not yet clear what the role of the Development Plan will be in the South African policy environment.

While South Africa has committed to taking 'nationally appropriate mitigation action' to enable its adherence to the PPD under the Copenhagen Accord, the term 'Nationally Appropriate Mitigation Action' (NAMA) is not extensively used in domestic climate mitigation policy dialogue. Notwithstanding this, South Africa is perceived as one of the most active countries on NAMAs in the international negotiations, having put forward suggestions on what NAMAs could look like, and how support for NAMAs could be organised through a registry linked to the UNFCCC mechanisms for finance and technology. The country's Copenhagen NAMA commitment is dependent on international support, but so far there are only initial attempts to secure this for specific activities (e.g. SARi). Some donor activity supporting activities to mitigate emissions are evident in the country, with a focus on pilots and capacity building; however, to date (August 2012) there has been no concerted or coherent effort by government, business or civil society.

4. A description of four South African mitigation actions

4.1. *Bus Rapid Transport in Cape Town*

The City of Cape Town's BRT system, known as 'MyCiti', is part of the Integrated Rapid Transit system envisaged as part of city-wide improvements of the public transport system. The BRT consists of trunk and feeder bus routes operating on dedicated bus lanes. The project comprises the physical infrastructure (roadways and stations), the buses, the centrally managed logistics control centre and tariff collection equipment (DME, 2009).

The MyCiti network will be rolled out over the next 15–25 years to serve the entire city. The implementation of the first phase 1a started in 2010 for the FIFA Soccer World Cup with an inner city loop service, and in May 2011 a second line was operationalised.

Although the BRT project is not specifically a climate-driven initiative, it has potential to reduce emissions through encouraging a modal shift from single occupancy car usage to buses, thereby providing gains in energy efficiency and reduction in fuel combustion. A project identification note (PIN) was prepared by the City of Cape Town as part of the initial stages of a CDM project, and identified that there is the potential to reduce 1.4 Mt CO_2/year (DME, 2009). At this stage the CDM route has not progressed beyond the PIN.

The City of Cape Town has been responsible for planning, designing and implementing the BRT with some additional input from external consultants, as this is the first Integrated transport system to be designed in South Africa. The operational aspects of running the BRT network such as vehicle operators, station management, a central control centre and fare management will be contracted out. The vehicle operating companies will be designed in such a way as to incorporate existing businesses currently operating on that route.

A component of the National Public Transport Strategy and Action Plan (DoT, 2007) includes implementing high-quality, integrated, mass rapid public transport networks (IRPTNs). The MyCiti project is part of the National Department of Transports IRPTN programme and is funded primarily through the Public Transport Infrastructure and System Grant (PTISG) (CCT, 2011). Furthermore, the 2009 National Land Transport Act (RSA, 2009) supports cities in developing their own public transport needs from planning aspects to the administering of funds for capital expenditure and subsidies. The Integrated Development Plan for the City of Cape Town has identified eight strategic focus areas of which improving public transport is one (CCT, 2007). This has been taken forward through the City of Cape Town's Integrated Transport Plan (CCT, 2007).

According to the Business Plan for MyCiti Phase 1a, the implementation costs are ZAR 4.6 billion. It is estimated that for Phase 1a the operating costs will have a deficit of ZAR 375 million between 2010 and 2014. The funding for this project comes from a variety of national- and city-level sources including the Public Transport Infrastructure Services Grant (PTISG), the City of Cape Town's Capital Replacement Reserve and the External Financing Fund, the Public Transport Operating Grant, as well as local rates, a share of the fuel levy, advertising and parking revenue. The largest source of funding is from the PTISG, a national grant promoting the provision of appropriate Integrated Rapid Public Transport Network services in major South African cities. The 2010 FIFA Soccer World Cup was a key driver for implementing the project. Leading up to the world cup there was significant national funding made available for improving public transport in the host cities.

This project is a mitigation activity that has already begun implementation, largely due to available national funding and the world cup. There are, however, risks and challenges as the project progresses, including how to address the financial deficit, strengthening institutional capacity to manage a centrally managed system, potential opposition from existing taxi and bus operators, and the complexity in managing fee collection and revenue distribution. Furthermore, as it is the first large scale integrated public transport system in South Africa, the City of Cape Town has to build capacity beyond planning and coordinating of a large infrastructure project but also the operational aspects.

4.2. *South African renewables initiative*

SARi was initiated in February 2010 as an inter-departmental initiative by the Department of Trade and Industry (DTI) and the Department of Public Enterprises to investigate ways of facilitating an accelerated scaling up of Renewable Energy production in South Africa (Zadek, Ritchken, Fakir, & Forstater, 2010). It has an ambitious stated aim of defining 'an industrial strategy for securing the economic gains from an ambitious program of renewables development, including financing and associated institutional arrangements that would not impose an unacceptable burden on South Africa's economy, public finances or citizens' (DTI, 2010a).

At national level, SARi vies for prominence within a relatively crowded but largely uncoordinated energy policy space. The 2010 Integrated Resource Plan, gazetted in May 2011, outlines a new build of 17.8 GW of renewable energy for electricity generation by 2030 which equates to 9% of the SA fuel mix for electricity in 2030 (South African Renewables Initiative, 2010). However, as the National Energy Regulator of South Africa may only issue generating capacity licences within the framework prescribed by the IRP (Ecologic Institute, 2010), without changes to the regulatory framework, the Initiative's ambitious scale-up target of between 10% and 15% renewable in the energy mix by 2013 is likely to be unattainable within the stricture of the existing 'new build' framework (DTI, 2010b).

In 2010, the South African cabinet approved the New Growth Path (NGP) which places job creation at the centre of government policy. SARi's modelling suggests that the proposed scale-up of renewables would produce 35–50,000 jobs, thus aligning it with one national priority. SARi is designated as part of a scaled-up green economy programme according to the DTI's Industrial Policy Action Plan (DTI, 2011).

Emission reductions associated with SARi that could be subject to Monitoring, Reporting and Verification (MRV) would not be produced directly by SARi as an 'entity', but by any renewable installations (wind farms etc.) built as a consequence of the success of SARi.

In 2011, the Initiative moved into the detailed design phase and was officially launched as a collaboration partnership at COP17 by the South African Government, the European Investment Bank and the Governments of Denmark, Germany, Norway and the UK. Institutionally, SARi is now co-chaired by the Department of Energy and the DTI (South African Renewables Initiative, 2010). It is worth noting that the key risks and requirements mirror each other: a robust and transparent institutional design and the consequent procurement process and the exercise of significant political will or 'buy-in' will facilitate SARi. Conversely, without these, there is a substantial risk of a lack of implementation.

4.3. *Carbon tax*

The National Treasury has signalled its intention to implement a carbon tax as an economic policy instrument to achieve greenhouse gas mitigation in South Africa (RSA, 2012; Treasury, 2006, 2011). The National Climate Change Response White Paper (Republic of South Africa, 2011a) also identified that carbon pricing will be part of the mitigation policy suite. At the time of writing, the carbon tax itself remains at an early stage of development. In December 2010, the Treasury released a Discussion Document outlining the rationale for implementing a tax in South Africa, but gives very few design indications. Following this, the Treasury convened a stakeholder engagement after comments on the Discussion Document were submitted, where modelling of the impact of the tax on the economy was presented. Preliminary design indications have been given in the 2012 Budget, together with a timeframe for the implementation of 2013/14, but there are still significant details outstanding.

While a carbon tax in itself does not reduce emissions, the changes in relative prices and responses that this induces have resulted in very significant mitigation in countries that have implemented carbon taxes (Winkler & Marquard, 2011). Academic work on a possible carbon tax in South Africa suggests a similar effect, inducing MAs of many different kinds across the economy (Goldblatt, 2010; Winkler & Marquard, 2011). Modelling for the LTMS analysed a carbon tax separately from other measures, in an energy model and also economy-wide modelling, and found that the carbon tax was the largest single mitigation intervention. Despite the indirect nature of its effects, a carbon tax is therefore a significant policy instrument for mitigation.

According to the Budget (RSA, 2012), the tax will have a broad base, covering a large number of South Africa's emissions sources and thereby potentially inducing significant mitigation. As the tax is likely to be levied on fossil fuels, and as supporting policies take effect, it may be difficult to identify the number of tonnes of CO_2 equivalent (tCO_2e) mitigated in response to the tax. If the carbon tax covers a large part of the economy, its mitigation benefits might be found through GHG inventories rather than through direct measurement of reductions. The tax may have significantly positive knock-on effects, as the price of carbon is anticipated to stimulate low-carbon industries, products and processes across a range of sectors. The Budget indicates a tax rate starting at R120 per tonne CO_2e, but subject to significant discounts. The tax itself as currently proposed (without any supporting incentives) will self-finance, and will generate revenue although this is not its primary purpose. How carbon tax revenues are recycled is crucial in analysing the broader socio-economic implications. Economy-wide modelling shows that revenues recycled in favour of poor households (e.g. through increasing the poverty tariff) can lead to overall positive impacts in those sectors (Kearney, 2008; Pauw, 2007). Another possibility is to use revenues for industries more sensitive to carbon pricing due to higher energy-intensity and providing incentives to improve energy efficiency. Both of these options are mooted in the Budget Review.

The carbon tax appears to be aligned both with national mitigation policy objectives and with the government priorities of growth and employment creation, depending, however, on how the revenues are recycled (National Treasury, 2011).

While Treasury is an influential department, and is determined to introduce the tax, it engages and potentially threatens incumbents in the minerals and energy sectors, and as such may face substantial opposition or lobbying from vested interests. The complexity of applying an economic instrument to an uncompetitive and regulated energy sector may also prove problematic. Institutionally, Treasury has the capacity to administer the tax, but may require assistance from external consultants and experts to design the tax, particularly those with an understanding of energy economics. While these are available domestically, international assistance is likely to be beneficial.

Therefore, the tax is an ambitious policy instrument, and one that is central to the country's mitigation efforts. However, it requires careful design to ensure minimal negative impact on economic growth, and careful political and process management to navigate vested interests in South Africa's carbon-intensive economy. Ensuring policy contextualisation, alignment and co-ordination with the Department of Environmental Affairs (DEA)-led overarching mitigation policy will be important.

4.4. *National Sustainable Settlements Facility*

The NSSF will administer financing, enabling the Department of Human Settlements (DHS) to increase the mandatory specifications of all new subsidised housing in South Africa to include solar water heaters and thermal performance improvements such as orientation, roof overhangs and insulating building materials.[2] It is a public facility, relying largely on a combination of international and domestic public funding. The NSSF is designed to earn income through an international carbon market mechanism such as the CDM.

The Facility has substantial co-benefits in the form of improved health and reduced energy bills for home occupants, access to improved energy service for both urban and rural communities, employment generation, air quality improvements and reduced requirements for national electricity generation at peak times, and is therefore fully aligned with national priorities. It has knock-on mitigation benefits through the stimulation of demand for energy-efficient interventions and awareness of mitigation issues in the low-income sector.

The NSSF was conceived and developed by South-SouthNorth Projects Africa (SSN), a Cape Town-based NGO. The SSN-initiated Kuyasa Housing Project demonstrated the use of sustainable energy technologies in low-income housing in South Africa, together with approaches for crediting emission reductions from these under the CDM (SSN, 2004), but did not consider financing for the project beyond the CDM. This becomes critical to achieve scale, hence the development of the NSSF. SSN has partnered with the Development Bank of Southern Africa, which will host the Facility, at least in the development phase. The DHS, Department of Energy (DoE), DEA, and Department of Science and Technology, together with the South African National Energy Research Institute (now called the South African Energy Development Institute), National Energy Efficiency Association and key Metropolitan councils have been involved in developing the concept and have given it their formal support. This formal support across several government departments makes this one of the more 'official' MAs of the four considered in this paper.

The NSSF is in the early design and planning stage, having been proven conceptually and subjected to extensive stakeholder engagement. Further development on MRV for the programme is being undertaken under the Gold Standard and the CDM,[3] with local technical capacity. The NSSF is being piloted at a large project scale in the housing development Cosmo City to develop and demonstrate a sustainable financial model.

The NSSF requires proving at scale, and detailed development of the mechanisms for implementation, including an institutional structure, disbursement mechanisms, MRV systems and access to the technologies and skills to install and maintain them. Proving a sustainable financing model, which effectively balances the interest and capabilities of all stakeholders, is the primary challenge of the NSSF. However, it also relies on many supportive measures that are not necessarily yet in place, including local skills to install and monitor the technologies, acceptance of the interventions by the households, a supply of domestically manufacturing technologies, and maintenance capabilities at scale. The current delivery of low-income housing is challenged by corruption and delivery issues on the ground, and overlaying a complex environmental mechanism on top of this may exacerbate these issues. However, it is also possible that these challenges could be overcome through design.

Emission reductions are achieved indirectly by the mechanism, through the housing projects which the NSSF financially enables. These are anticipated to be in the region of 25 Mt over the first 10 years of the project, at a cost of around ZAR370 per tonne. This figure excludes energy saving and co-benefits. Depending on the eventual MRV requirements, a capacity to MRV may exist in the country, but is likely to require further development.

The NSSF needs to gain traction politically and momentum through confidence-building examples and pilots. As it spans the Housing, Energy and Environment sectors, some level of co-operation and co-ordination of these three departments on the NSSF would be advantageous.

5. Analysis of the examples

A set of criteria was developed in order to consistently analyse the four examples against the study's three areas of enquiry. The criteria were developed by firstly drawing from previous MAPS work[4] on NAMAs as well as existing sources from international think-tanks like Ecofys, before posing ourselves a range of questions to broaden our understanding of how MAs are or might be defined, supported and potentially implemented in South Africa.[5]

The resultant criteria fall into three subsets: descriptive, implementation issues, and NAMA-specific elements with a view to addressing issues of international financing and reporting. These are presented in the tables appearing in this section, which summarise each example's performance against the criteria and provide the basis for the authors' analysis.

The analysis is admittedly limited in that it considers only a very small and randomly chosen sub-set of South African MAs, all but one of which have not yet been implemented. However, it raises a number of issues relating to the South African approach to MAs, barriers to implementation and how implementation can be supported going forward.

5.1. *Describing the examples*

The four examples could be categorised as a local transport project; a national financing mechanism for renewable energy; a national mechanism for financing, aggregating and facilitating sustainable energy interventions in the low-cost housing sector; and an economic policy instrument (carbon tax). Therefore, while each covers different areas of mitigation, there is a strong weighting towards those that are focused on financing in the sample.

Of the four examples examined, only the BRT has been implemented, and then only in its first phase. The tax and SARi are likely to be implemented in the medium term (within two to three years) while the NSSF is a long-term mitigation project. It is interesting that only the BRT – whose primary motivator was not climate mitigation – has 'broken ground'.

All of the activities to mitigate emissions can be relatively easily modelled to understand their emission reduction potential. However, this does not necessarily reveal much about the issue of attribution, particularly in the case of the carbon tax and SARi. For example, whilst SARi would primarily provide coordination of more ambitious action on renewable energy and a proposed financial framework to encourage investment, any reductions would not be produced directly by SARi as an 'entity', but by any renewable installations (wind farms etc.) that were built as the consequence of the success of SARi. Emission reductions could therefore only be counted either for individual renewable energy projects or for a programme like SARi as a whole, but not for both.

The examples were further considered for their ability to produce 'knock on mitigation effects', or mitigation beyond the individual activity's boundaries. This varied amongst the examples, pointing perhaps to the difference between local and national level activities to mitigate emissions, and also potentially to the difference in the number of people, economic actors, end users or sectors the activity reached. The type of activity may also play a role, with financial mechanisms having a greater potential for additional mitigation, although potentially for different reasons (Table 1).

The four examples provided an opportunity to explore a South African approach to MA, and to gain an insight into some of the challenges of generating an MA definition. Attribution is a key element in determining an MA, but it is not always straightforward. Whilst projects and programmes which directly reduce emissions are uncontroversially MAs, those that act as indirect levers or incentives, such as the carbon tax, or financing initiatives are less clear-cut. Even more controversial are information initiatives, enabling activities and general policies or strategies. Further issues raised but not resolved by the authors included: whether it is useful to have a definition of MA at all; the importance of an implementation focus to any MA definition; and that the reason for establishing a definition would impact the outcome. Further discussion on what was agreed by the authors to be a side issue to this paper can be found in the extended report from which this analysis is drawn (Tyler et al., 2011).

5.2. *Implementation issues*

The analysis presented in this and the following sub-section is based on the authors' assessment of each MA's level of risk against the criteria. A one-star (low risk), two-star (medium risk) and three-star (high risk) system of risk or feasibility ranking is applied.

The application of the implementation risk criteria to the four examples was a challenging and fundamentally subjective process, given that most of the MAs have not yet been implemented. In addition, because the examples are at different levels of maturity, the analysis focuses on areas which have been challenging to implementation,

Table 1. Describing the examples.

Criteria	Timing: is this a short, medium or long term MA?	What are the relative CO_2e saving?	Potential mitigation knock on effects (S, M, L)	Any co-benefits?
Tax	M	No figures given but could be deduced	L (behaviour, price internalisation)	Recycling or revenue may yield co-benefits
NSSF	L	6 Mt per annum at full operation	L (education, manufacturing industries)	Skills, employment, health, energy poverty alleviation, avoided electricity generation
SARi	M	1.2 Bt tonnes by 2045 or 60 Mt per annum at full ramp-up	M (increasing of manufacturing base etc.)	Manufacturing, air quality, FDI attraction, skills development
BRT	S	1.4 Mt of CO_2e over the first 10 years of implementation	S (awareness, increasing capacity at municipal level)	Improved air quality, reduction in transport costs, avoided fuel consumption, BoP benefits. Develop local construction skills base, formalise and grow taxi industry

Note: FDI: foreign direct investment; BoP: balance of payments.

rather than assessing current levels of implementation. The analysis is presented in Table 2.

The highest level of risk to all the examples can be found in the two right-hand columns 'vested interest opposition' and 'additional financing requirements'. International climate finance is largely being identified to cover the additional financing requirements. It is interesting that these two criteria are not yet present in the international literature relating to NAMAs. Overall, the carbon tax comes out as the least risky MA, with only technical capacity to operationalise and vested interests a potential risk. The NSSF appears to be the most risky, followed by SARi and then the BRT.

Risks arising from a poor mandate and weak or poorly defined ownership for the MA could occur when there is a disparity between what the owner typically does, and what the MA does (in the case of the NSSF), or when the MA cuts across government departments (e.g. SARi which appears to straddle Departments of Public Enterprise, Trade and Industry and Energy, and the NSSF which straddles both the DHS and the DoE). Clear, high-level and possibly legal mandates for MAs would facilitate implementation.

The risk of insufficient institutional capacity to take the MA to implementation seems to diminish, the closer the MA is aligned to successful mainstream and existing activities (e.g. the tax, which is a variant on the very well established policy tool of taxation, appears to have a lower implementation risk than the NSSF, which is a first of its kind from a variety of perspectives).

A supportive policy, regulatory and planning context is an important enabling factor for implementation, and the risks of this seemed to diminish where there is a non-climate change mitigation driver (such as the world cup driver for the BRT), or a strong owner (such as the Treasury for the tax). As noted when considering the risk of insufficient mandate and poorly defined ownership, MAs which need to cut across government departments may encounter greater challenges (eg SARi, NSSF). One way around this may be to situate an MA clearly in one area (eg the BRT in transport), with co-benefits in another (energy efficiency), but not attempt to require both to drive the MA.

Interestingly, all four examples are aligned to national priorities. This should help with implementation, and tends to suggest that they are nationally appropriate and that their mandate could be part of broader mandates or enhancements of existing mandates.

A financial structure was completed for the BRT (first phase) and SARi. The NSSF's financial structure is under development, and the issue of a financial structure is not

Table 2. Enabling implementation.

Descriptive criteria	Tax	NSSF	SARi	BRT
Is there a problem with mandate?	⊗	⊗⊗	⊗⊗	⊗
Is there relevant existing institutional capacity to implement?	⊗	⊗⊗	⊗⊗	⊗⊗
Is there a supportive planning, policy and regulatory context for the MA?	⊗	⊗⊗⊗	⊗⊗	⊗
Is it aligned with national priorities?	⊗	⊗	⊗	⊗
Has a financial structure for the MA been developed?	⊗	⊗⊗	⊗⊗⊗	⊗
Is there local technical capacity to design the MA?	⊗ ⊗	⊗	⊗	⊗⊗
Capacity to technically operationalise MA	⊗	⊗⊗	⊗⊗	⊗⊗
Vested interests	⊗⊗⊗	⊗	⊗⊗⊗	⊗⊗⊗
Additional financing	⊗	⊗⊗⊗	⊗⊗⊗	⊗⊗⊗

particularly relevant for the tax which itself generates revenue and does not require much financing to implement. Importantly however, these are still at a modelling phase, and the only implemented MA, the BRT, is still in deficit. Importantly, this criteria also does not consider whether the MAs are financially viable (captured in the 'additional financing' risk category).

Local capacity does exist to design the NSSF and SARi, but the BRT required external expertise, and the design of the tax may too. Interestingly, it appears that the MAs which are more unique to climate mitigation (SARi and the NSSF), and further from mainstream approaches (the BRT and tax), may struggle less with capacity issues, suggesting a high level of local innovation on the design of activities to mitigate emissions. These are also the MAs which are most advanced in their development as NAMAs.

Technical capacity to operationalise was a risk to the BRT, but it is not known to what extent this may be problematic for the remaining MAs. However, any new initiative (mitigation or other) inherently requires support. If the underlying area is well functioning (e.g. taxation), then this may raise the likelihood of the MA being successfully implemented. In turn this suggests that South Africa should prioritise MAs which are well aligned to the country's strengths.

Therefore, aspects which are playing an important role in the development and implementation of some MAs, and may be able to be expanded to others, may include:

- Alignment of the MA with underlying areas of national strength in implementation and strong mandate, particularly to counteract the present bias against implementation overall
- Close alignment of the MA with the core business of the owner or implementer
- Exploiting the use of local development co-benefits which are likely to help achieve a mandate to implement. A tenuous finding is that MAs driven by non-climate factors may have a greater chance of implementation. This could possibly be translated as 'where-ever possible drive through other avenues, or link closely to other motivators for the project'.

While challenges to MA development and implementation could be counted as:

- MAs struggle with typical project development blockages.
- Financing and vested interests are identified as large constraints to successful implementation.
- Technical capacity to design and operate MAs is required.
- A final constraint identified is the weakness of many state institutions which give effect to delivery.

Some of these challenges are likely to be best addressed domestically, while others will require international support. For example, financing and technological capacity are something that international support will be able to assist with, but countering vested interests is largely a domestic issue, and specific to each MA. Typical project development blockages, and the weakness of existing state institutions which give effect to delivery are more intractable issues, and it is less clear how the international community could assist in overcoming these.

5.3. *NAMA-specific elements*

The NAMA route currently appears to be one of the most likely mechanisms through which MAs could receive international assistance, and each example was considered against criteria which are considered important for NAMA suitability. The results are reflected in Table 3 (the star system is partially utilised here again).

It would appear from the table that it would be advantageous to promote both SARi and the NSSF as NAMAs, primarily due to the level of incremental financing required to implement them. In the case of the BRT, it is less easy to measure attributable emission reductions. It is not clear how the tax could be promoted as a NAMA, despite its anticipated significant contribution to South Africa meeting its Copenhagen Pledge.

The level of incremental costs is very difficult to determine for the BRT, and potentially too for other areas where infrastructure is going to be delivered anyway. For the other three examples, the incremental costs are likely to be

Table 3. Supporting implementation: NAMA-specific criteria.

Criteria	Tax	NSSF	SARi	BRT
What is the level of incremental cost required to implement?	0	ZAR 5 billion	ZAR 74 billion adjusted	Phase 1A – Implementation costs are ZAR 4.6 billion
What is the full development cost?	–	ZAR 10 million	unclear	Operational ZAR 375 million between 2010 and 2014
Is this MA being designed to receive climate funding?	No	Yes	Yes	No
Can the emissions be monitored, reported and verified?	⊗⊗	⊗	⊗	⊗
Do we have the capacity to monitor, report and verify emissions	⊗	⊗⊗	⊗	⊗

available. Estimating the costs up to the point of implementation is anticipated to be less complex than estimating the full cost of implementation.

SARi and the NSSF have been specifically designed in order to receive climate funding, while the tax and the BRT have not. This may be due to the underlying nature of the MAs themselves or the fact that the more mitigation specific and innovative the MA is, the more likely it will have mitigation financing as a focus.

All of the four examples will be able to provide information that would be needed for MRV of actions, although this may be difficult for the tax due to the way it is proposed to be levied (on fossil fuels as an indirect proxy). The nature of the MRV is very project specific, and it must be designed in a way that is appropriate and enabling. The level of detail on financing suggests that these MAs would also be able to quantify the financial support required for implementation.

As a cautionary note, just because it may be easy to develop a NAMA in a particular area, this does not necessarily mean that it is the right thing for South Africa to do. South African should focus first on developing MAs that are inline with their national development priorities and domestic capacity, and secondly on whether they then fit an emerging international support mechanism.

6. Findings and conclusion

A consideration of South Africa's mitigation context and policy approach to MAs reveals a range of approaches to implementing and finding support for individual MAs and broader enabling activities. However, there are many diverse MA activities under way, and the policy context is evolving. MAs currently under development experience a range of challenges, only some of which are likely to be able to be addressed with international support. Implications for the developing NAMA mechanism could be drawn from this.

In order to gain additional understanding of MAs in South Africa, a bottom-up analysis of four examples of activities to mitigate emissions were analysed according to a set of criteria developed for the purposes of this analysis. Probably most significantly, each MA considered was found to be very different, suggesting that the remaining MA population may be equally diverse. If this is confirmed, it would imply that caution should be exercised in seeking standardisation of MAs and by implication, NAMAs. The case study approach, while limited in the extent to which its findings can be generalised, does shed some light on understanding MA from a bottom-up South African perspective.

Arriving at a definition of an MA proved difficult, and was resolved. A number of questions arose which may be worth taking up in future work.

The only MA of the four examples considered in the paper which had been (partially) implemented at the time of writing, the BRT, was not driven by mitigation concerns at all, but rather by the availability of funding for the 2010 FIFA Soccer World Cup. A number of factors appear to be constraining MA implementation, including the availability of financing, vested interests, and a weakness of state institutions. Some of these are able to be addressed through international support, while others are inherently domestic issues that could possibly be countered by aligning MAs with current areas of implementation strength within South Africa, focusing on a broader set of national development priorities and avoiding MAs that fall across more than one government department. A unified understanding of MA, and a plan or series of plans to implement MAs would clearly be beneficial, and all the more so if the mandates for action are linked to actual projects and programmes that directly deliver emission reductions.

With regard to receiving international support, certain MAs appear more appropriate for development as NAMAs, because they clearly require significant financing and because they lend themselves to MRV. It is important that MAs drive the development of NAMAs, not the other way round, as this may lead to problems with implementation if the NAMAs are not aligned with areas of implementation strength in the country.

Further work is required to verify these findings, and to expand on them. At this point, it is hoped that they will catalyse discussion within the MAPS programme and beyond regarding approaches to, implementation of and support for MAs.

The study also raised a number of areas for further research. Analysis of a broader set of MAs, particularly including a greater variety of types, could enrich the working definition suggested here and aid implementation. Each criterion used in the analysis could also be significantly further defined, most usefully in conjunction with an interrogation of a broader set of MAs. The role of the private sector in South African MAs could be further explored. Finally, an analysis of additional implemented MAs, where available, would strengthen the findings.

Notes

1. The MAPS programme is a collaboration between a number of developing countries, promoting best practice in mitigation action planning or scenarios development. The MAPS programme seeks to achieve this by support to in-country processes informed by research. It seeks to share and deepen the leanings through collaboration between Southern experts supporting government programmes: http://www.mapsprogramme.org
2. While sustainable energy interventions are encouraged in the National Housing Code (2009), there is no corresponding budget allocated to finance these.
3. Including GS small-scale PoA registration, and continued work on large-scale methodologies which incorporate a 'suppressed demand' approach to crediting, and simplified monitoring requirements.

4. Further details of the development of the criteria is available in the full report: Tyler, Boyd, Coetzee, and Winkler (2011)
5. Further details of the development of the criteria is available in the full report: Tyler, Boyd, Coetzee, and Winkler (2011).

References

City of Cape Town (CCT). (2007). *5 Year plan for Cape Town. Integrated Development Plan (IDP) 2007/8–2011/12*. Cape Town: City of Cape Town. Retrieved from http://www.capetown.gov.za/en/IDP/Documents/IDP_review_elephant_Jun_08_web.pdf

City of Cape Town (CCT). (2011). *Project Status and Progress Report No 14, March 2011 Progress Report from March 2011*. Retrieved from http://www.capetown.gov.za/en/MyCiti/Pages/Monthlyprojectreports.aspx

Climate Analysis Indicators Tool (CAIT). (2011). Version 8.0. Washington, DC: World Resources Institute. Retrieved from http://cait.wri.org

Department of Minerals and Energy (DME). (2009). *Clean Development Mechanism South Africa Designated National Authority*. Project Identification Note, June 2009. Pretoria: Government Printer.

Department of Trade and Industry. (2010a). *Unlocking South Africa's Green Growth Potential*. South African Renewables Initiative (SARi) Update briefing, December 2010. Retrieved from http://sarenewablesinitiative.files.wordpress.com/2011/02/sari-cancun-update-briefing-for-distribution.pdf

Department of Trade and Industry. (2010b). *Unlocking South Africa's Green Growth Potential*. South African Renewables Initiative (SARi) Update briefing presentation at COP16, December 2010. Retrieved from http://sarenewables initiative.wordpress.com/about/the-sari-approach/

Department of Trade and Industry (DTI). (2011). *Industrial Policy Action Plan (IPAP), 2011/2012–2013/14, Economic Sectors and Employment Cluster*. Pretoria: Government Printer.

Department of Transport (DoT). (2007). *Public Transport Action Plan Phase 1 (2007–2010): Catalytic Integrated Rapid Public Transport Network Projects*, February 2007. Pretoria: Government Printer.

Ecologic Institute. (2010). *Financing needs and preconditions for the South African Renewables Initiative (SARi)* (Final Report). Cape Town: Killian Wentrup, Ecologic author.

Goldblatt, M. (2010). A comparison of emission trading and carbon taxation as carbon mitigation options for South Africa. *Climate Policy*, *10*(5), 511–526.

Kearney, M. (2008). *Long-term mitigation scenarios: Dynamic economy-wide modelling*. Technical discussion document. Cape Town: Energy Research Centre.

National Treasury. (2006). *A framework for considering market based instruments to support environmental fiscal reform in South Africa*. Pretoria: Tax Policy Chief Directorate.

National Treasury, Chief Directorate: Modelling and Forecasting. (2011). *Economic impacts of introducing a carbon tax in South Africa: Provisional results*. Presented at the carbon tax stakeholder workshop, March 2011. Retrieved from http://www.treasury.gov.za/divisions/tfsie/tax/CarbonTaxWorkshop/default.aspx

Pacala, S., & Socolow, R.H. (2004). Stabilization wedges: Solving the climate problem for the next 50 years with current technologies. *Science*, *305*, 968–972.

Pauw, K. (2007). *Economy-wide Modeling: An input into the Long Term Mitigation Scenarios process* (LTMS Input Report 4). Energy Research Centre, Cape Town, October 2007.

Republic of South Africa (RSA) (2009). National Land Transport Act, 2009 (Act No. 5 of 2009).

Republic of South Africa. (2010). *Letter to the Executive Secretary of the United Nations Framework Convention on Climate Change (UNFCCC) on South Africa's nationally appropriate mitigation action under the Copenhagen Accord*. Retrieved from http://unfccc.int/files/meetings/cop_15/copenhagen_accord/application/pdf/southafricacphaccord2_app2.pdf

Republic of South Africa. (2011a). Pretoria: Department of Environmental Affairs, National Climate Change Response White paper 2011.

Republic of South Africa. (2011b). Electricity Regulation Act (4/2006): Electricity Regulations on the Integrated Resource Plan 2010–2030 (Government Notice No. R. 400, 2011) Government Gazette 34263:551, May 6 (Regulation Gazette No. 9531), Pretoria.

Republic of South Africa., National Treasury. (2012). *National Budget 2012*. Retrieved from http://www.treasury.gov.za/documents/national%20budget/2012/default.aspx

South African Renewables Initiative. (2010). Unlocking South Africa's Green Growth Potential. Pretoria: Department of trade and Industry, republic of South Africa. Retrieved from http://sarenewablesinitiative.wordpress.com/analysis/

SouthSouthNorth. (2004). *Kuyasa Low Cost Urban Housing Energy Upgrade Project: Project Design Document*. Retrieved from http://cdm.unfccc.int/Projects/DB/DNV-CUK1121165382.34/view

Tyler, E., Boyd, F., Coetzee, K., & Winkler, H. (2011). *Country study of South African Mitigation Actions*. Research Report for MAPS (Mitigation Action Plans and Scenarios). Cape Town: Energy Research Centre, University of Cape Town. Retrieved from http://www.mapsprogramme.org

Winkler, H., & Marquard, A. (2011). Analysis of the economic implications of a carbon tax. *Journal of Energy in Southern Africa*, *22*(1), 55–68.

Zadek, S., Ritchken, E., Fakir, S., & Forstater, M. (2010). *Developing South Africa's Economic Policies for a Low Carbon World*. Retrieved from www.zadek.net

Comparative analysis of five case studies: commonalities and differences in approaches to mitigation actions in five developing countries

José Alberto Garibaldi[a], Harald Winkler[b], Emilio Lèbre La Rovere[c], Angela Cadena[d], Rodrigo Palma[e], José Eduardo Sanhueza[f], Emily Tyler[g] and Marta Torres Gunfaus[b]

[a]Independent Climate Consultant, London, UK; [b]Energy Research Centre, University of Cape Town, South Africa; [c]Center for Integrated Studies on Climate Change and the Environment, Institute for Research and Postgraduate Studies of Engineering, Federal University of Rio de Janeiro, Brazil; [d]Universidad de los Andes, Bogotá, Colombia; [e]Universidad de Chile, Santiago de Chile, Chile; [f]Cambio Climático & Desarrollo Consultores, Santiago de Chile, Chile; [g]Independent Climate Economist, Cape Town, South Africa

In light of the ongoing international discussions about the Nationally Appropriate Mitigation Action concept, this study takes instead a more 'bottom-up' approach through a comparative analysis of five studies of mitigation actions (MAs) in Brazil, Colombia, Chile, Peru and South Africa. The analysis shows that MAs are driven by both developmental and climate objectives. The character, scope, policy horizon and potential success of an action are closely linked to the developmental path of countries such that MAs that directly address poverty and development seem to have a better chance of being implemented since they address issues higher on the policy agenda of developing countries. Where international support is sought, all five countries have some existing measurement, reporting and verification (MRV) and technical competence capacity that can be built upon. The choice of MAs is evidently linked to institutional capacity (both for design and implementation of MAs and possible MRV), emissions profile and the relative resource endowments of countries. The policy environments – from highly planned to less coordinated – and time-horizons – from 4-year plans to 40-year scenarios – differ substantially between the countries. Thus, the comparative analysis underscores the diversity of possible MAs and capabilities and the concomitant need for flexibility in definition, design and implementation.

1. Introduction

The objective of this study is to develop a better conceptual understanding of mitigation action (MA) by comparing countries' approaches to thinking and implementing MAs. The comparative analysis in this study builds on the textured and detailed analysis of MAs in five developing countries based on the articles on this volume – Brazil, Chile, Colombia, Peru and South Africa (Cadena Monroy et al., 2011; La Rovere, Dubeux, Perira, & Wills, 2011; Sanhueza & Palma, 2011; Takahashi, Zevallos, & Cigaran, 2011; Tyler, Boyd, Coetzee, & Winkler, 2011). The reader is referred to these studies for further detail; to avoid repetitive citations, we cross-reference only sparingly.

In this comparative analysis, we assess what is common across MAs in all contexts, and what is different. We structure the comparison by considering the elements specified in the methodology in Section 2. Thus, in the subsequent sections, we will address some key concepts, advance a methodology, provide a detailed comparison across the five countries for key elements (Section 4) and synthesize this in a summary comparative table in order to finally draw conclusions.

2. Some key concepts: what is in a name

There are several ways to refer to the relevant concepts. This paper will focus on MAs. We will consider these as actions that result in the mitigation of GHG, as they are on-going in all of the compared countries, even if these were deployed for reasons of local sustainable development, rather than for climate change purposes. In this sense, we distinguish MAs on one side from low-carbon development strategies (LCDS) – countries put together (see Torres, Winkler, Tyler, Coetzee, & Boyd, 2012; UNEP, 2011). In contrast, from Nationally Appropriate Mitigation Actions (NAMAs), as those tend to be associated under the international climate negotiation as individual actions to be submitted to the registry under the United Nations Framework Convention on Climate Change (UNFCCC). NAMAs are being linked by the climate regime to a series of support institutions as they are being developed at the UNFCCC. MAs developed domestically may be submitted internationally as NAMAs, particularly if international support is sought for their implementation.

We believe that it is most important to understand the country's approaches to MAs in order to gain broad support and increase the number and ambition of actions implemented. A key challenge in developing countries is to get from policies and plans to implementation, given resource, institutional and capacity constraints. Moreover, in assessing what is common across MAs in all context and what is different or context-specific, this study strikes a balance between a wider and narrow view – more gets to be seen and more opportunities arise, while allowing addressing other key policy and national concerns, such as poverty and development. Hopefully, this will help identify what implicit and tacit knowledge we already have, what alternatives there are and how to better use them.

3. Methodology: how will we compare – and some guidelines

The comparison is advanced as arising from the relevant issues considered in the country case. We assess the MAs from the five countries against the following elements: conceptual approaches, including planning and regulatory concepts, the stage of development of MAs (including specific examples), capacity (both institutional and technical), poverty and development, ownership and finance. Section 4 is structured around these elements. In assessing a diversity of MAs from five different countries, we are not suggesting a common template, but using these elements for comparative analysis, that is to identify similarities and differences.

These categories help focus the analysis on the capacity for policy action of domestic agents and institutions, taking into account the resource base and emissions profile within which they operate and which affects them. Thus, it can consider both how these agents can imagine different development futures as well as their capacity to act and implement the changes required to bring them forward – considering their circumstances. In doing this, the countries' own domestic options and constraints can be assessed, without hindrance from those deriving from the international climate regime.

Moreover, these categories also facilitate the consideration of MA within larger emerging economies (i.e. Brazil and South Africa) as well as smaller, high growth emerging ones (i.e. Colombia, Chile and Peru). Their various challenges may differ: while in the former, larger emissions profile in many sectors might make it crucial to reduce emissions or their growth, the latter needs to avoid their future growth. A focus on MA allows exploring how underlying factors affect the required capacity to achieve this.

Last but not least, while linking well with the categories outlined above in examining the various countries' approaches, they also match well with UNFCCC negotiation language, and provide an adequate blend of guidance and flexibility to encompass both individual NAMAs and the more general LCDS as required. A focus on the latter alone would leave aside the potentially larger set of MAs the country is advancing on its own for whatever reason, focusing instead on the more limited number of actions that seek support and/or have been deployed following the UNFCCC's or best practice NAMAs or LCDS guidelines. As not all MA needs to eventually translate into either NAMAS and/or LCDS, a focus on the latter can be a serious bias.

Thus, an interpretative guideline would take cue from the policies and activities described in the context and circumstances surrounding the country case articles, and not solely on *ad hoc* NAMAs. It would also consider a closer link to the relation between the country's general policy framework and its mitigation opportunities and challenges, and would follow climate change mitigation approaches that shadow the main policy interests of the country, whether or not they emerge from the international climate regime.

4. Detailed comparison of MAs by issue

Having considered the key concepts and outlined our methodology, this section provides a detailed analysis of MAs across key issues, identified in Section 2. The main sources for this analysis are the five studies from researchers. Readers wishing to see the comparison in tabular format might wish to look ahead to Table 1 at the end of this section.

4.1. *Concept*

The countries in all the cases examined in this article developed MAs initially as part or with a component including the Clean Development Mechanism or as autonomous developmental actions. In all five countries, MAs (and NAMAs particularly) were subsequently conceptualized more explicitly as part of pledges made in advance or in the context of Copenhagen (2009), at the fifteenth Conference of the Parties (COP-15) to the UNFCCC and the meeting of parties to the Kyoto Protocol. These were early movers with relatively ambitious overall MA goals – the smaller countries moving first, but all seeking to make an impact in the post-2012 negotiations.

In the opening of the high-level segment of COP-14 in Poznan in 2008, Peru offered substantial MA in forestry in exchange for further action by the developed countries. The same year, South Africa presented in a side event its long-

Table 1. Synthesis of comparative analysis of mitigation actions in Brazil, Chile, Colombia, Peru and South Africa.

	Brazil	Colombia	Chile	Peru	South Africa
Concept	National-wide target compiled in Law. Focus on AFOLU and energy	Sectoral targets embedded within national plan. Focus on forestry, energy, biofuels and carbon markets	National-wide target and several NAMAs under development, focusing on energy, transport and forestry	Sectoral targets, focusing on forestry, energy and waste	National-wide target and planning several MAs including both direct actions, indirect instruments and institutions
Stage of development	Sectoral plan for energy, forestry and agriculture available, others under development as per Decree, with further actions to be compiled	Specific climate change plan and strategy collecting activities currently under discussion. A number of actions at forefront (energy incentives, forest certificates, BRT, scrappage and modal shift actions, biofuels), some with challenges ahead	Identified and being designed in energy and transport; preliminary work in forestry. They are mostly at a design stage	Suite of actions in the energy, industry, transport, forestry and agriculture and waste sectors, at different stages	Most in design stage, with some advances on the BRT, a pilot in the settlement facility for housing
Institutional capacity	MA collected and implemented by descentralized bodies. Implementation in charge on national entities; with law enforcement and policy setting in Brazilian hands, but financial support admitted with foreign funds, inclusive of 1BnUS$ Amazon fun, and with participation of subnational governments. Where relevant (particularly on agriculture)	Energy with capacity to design, implement and supervise; forestry likewise, but facing substantial inherent difficulties in deforestation; well structured scheme in biofuels, regulated by the MoA, and with close links with regional research bodies in agriculture	Ministry of energy with *ad hoc* bodies and expertise; and transport and forestry lacking these. However, the latter have long standing technical expertise. Forestry (under Agriculture) has the capacity to define and design NAMAS, while transport can do so with Ministry of environment support	Stronger in energy, less in industry, transport and waste. Forestry with some overlap between agriculture and environment, creating some conflicts of competence. This, together with a poorly structured civil service regime, high personnel rotation and relatively low salaries conspire against stability and follow up of proposals	Capacity will need to grow at city government while better combining with central government capacity
Planning, policy and regulatory context;	Planning done by ministries with input from technical bodies operating within sectoral legislation	Administrative and incentive control measures deployed by MinEnv. as well as MINAg and strong links with planning. Reliability on market, links to private sector and engagement with cities (transport)	Free market in energy, and several supervisory and regulatory in the forestry sector, lacking institutional definition	National mitigation guidelines available (consulted at regional level). Variations on implementation within sectors, with several levels and agencies of government participating	Included within national planning and city planning documents, with limited stakeholder consultation. Indirect instruments based upon existing institutions

(*Continued*)

Table 1. Continued.

	Brazil	Colombia	Chile	Peru	South Africa
Technical capacity to design MAs	Already designed; implementation through national fora; new MAs in course of being agreed; direct and indirect policy mitigation instruments and vehicles	Available in energy, transport, forestry, biofuels, and agriculture	Available in energy; less in forestry and last transport; but latter capable with support from environment	Most currently developed through public private partnerships. Some capacity in ministry of energy, but additional support required in other sectors	Existing both within different levels of government and civil society. However, some of it (e.g. at city level) might need to be expanded. Strong design capacity in central government (finance and energy) but potential opposition from strong entrenched vested interest
Poverty implications	Substantial in credits in agricultural case, indirect in forestry, and with impacts on urban conditions as well in the case of energy	Strong in deforestation and agricultural, indirect in transport, some in energy and biofuels	Some impacts in energy and forestry	Strong in forestry and agriculture, indirect in energy and transport	Strong and direct in urban housing project, indirect in the others
Ownership (who initiated and 'owns' the MA)	Public players, with private sector participation in implementation	Government entities (national and subnational –city governments) with some public private partnership	Initiated by government, and with Government ownership. Role for public private partnerships expected, although yet to materialize	Mix of government and consultants, with some regional government participation, particularly in REDD+	Mix of central and city government, with also strong participation by civil society and private sector, within a fluid, rich but still yet to be coordinated environment
Financing	Distinction between enforcement and legal and policy definition to support financing without compromise of national position. Flows through development Bank (BNDES), and finance blended with carbon markets and national support	Implicit private sector incentives through firm capacity charges, additional financing options possible; in transport, multiple options operative with others in design (scrappage and retrofitting), plus incentives in forestry. Additional action potentially viable through international support	In energy, some design of specific instruments (revolving funds, concessionary finance, subsides and credits, etc.). Costs estimated for forestry and energy, but not agreed framework in case of forestry yet	Strong potential for energy infrastructure activities to be privately financed; while in efficiency activities and those in other sectors there might be a need to blending financial sources and carbon finance. Additional support for technical expertise might be required	Several self-funding and innovative financial mechanisms (e.g. settlement facility and feed in tariffs) with clear overview of total costs, and some piloting experience. Also, a well honed ability to make the most of circumstances (e.g. using the World Cup finance)
Any other issues	Strong concern over ownership and sovereignity in case of forestry sorted through MA design	Some overlapping of competences, in agriculture and forestry; none in case of alimentary security and			

Data source: Analysis by authors, based on own research and information in the studies by (Cadena Monroy et al., 2011; La Rovere et al., 2011; Sanhueza & Palma, 2011; Takahashi et al., 2011; Tyler et al., 2011).

term mitigation scenarios (Republic of South Africa (RSA), 2008), but still informally and not as a matter of negotiation. The Peruvian case emerged from a specific analysis of what it gained and lost by being more or less ambitious in a high ambition international coalition, taking into account mitigation, adaptation and impacts (Ministerio del Ambiente, 2010). Its 2010 national mitigation guidelines explicitly argued that the country would be better off in a high ambition outcome towards which it contributed much, than in one of the low ambition in which it had to little or no MA committed.

In the lead-up to COP-15 in Copenhagen, all four had followed Peru in formally announcing pledges. Chile, Colombia and Peru had exchanged views along the lines above on mitigation and on the interaction between domestic policy and the international climate regime regularly since 2006, within the Latin American Workshop to increase the scale of responses. With variations, the first three sought to respond to the lack of international action in the run up to Copenhagen and its aftermath, using domestic high ambition mitigation pledges to elicit further international collective action (Garibaldi, Araya, & Edwards, 2012). They pledged their actions after the Copenhagen COP.

Brazil and South Africa were in a different setting, and their goals differed: Brazil's 'offer' specified individual MAs, South Africa's did not. In spite of these differences, Chile and Brazil remain some of the few developing countries that have *both* an overall deviation/intensity goal, and specify that these will be achieved by specified MAs. Indeed, this is more specific than required for developed countries under the Kyoto Protocol, where the commitment is quantified overall, but policies and measures are flexible. Among the countries studied, there are two economy wide goals (South Africa, Brazil, both with quantified deviation below business-as-usual (BAU)). From the cases, it seems clear that all offers were affected by the countries' emissions profile and their respective resource endowments – an issue we return to in Section 5.

Brazil's MAs have been encoded in law and implementation decrees, covering forestry, agriculture and animal husbandry and energy. Goals were quite ambitious, reflecting proposals for 36.1 and 38.9% reduction for 2020 (La Rovere et al., 2011).

South Africa formally communicated its pledge to the UNFCCC after Copenhagen, indicating that its NAMA would enable a 34% deviation below BAU to 2020 and 42% by 2025, with the extent to which this action will be implemented dependent on support (RSA, 2010). South Africa now has a desired GHG emission trajectory, and has also specified a BAU, both described in its climate policy (RSA, 2011). MAs under consideration to achieve the deviation include both direct actions, indirect instruments and institutions – from carbon taxes and tariff regulation and incentives to settlement funds and bus rapid transits (BRTs) – at different points in the development and design of the action, all these are described within.

Chile's goal is to reduce the growth of CO_2 emissions by 20% of the BAU scenario by 2020, using 2007 as a base year, provided international assistance is available (Sanhueza & Palma, 2011). Several NAMA options are described in the pledge: Five in Energy, four in transport and two more in forestry. The latter offers substantial mitigation opportunities (over 230 Mt of CO_2-eq) compared to some 10 Mt in the energy sector.

Peru has put forward three major MAs in its pledge, including achieving net zero deforestation of primary natural forests by the year 2021 and stabilizing by 2017 emissions from protected areas. Another MA aims to ensure that by 2020, renewable energies (nonconventional, hydropower and biofuels) make up at least 40% of energy consumption, emissions from protected areas. The third MA involves designing and implementing measures that reduce emissions from inadequate solid waste management (Takahashi et al., 2011).

Colombia offered zero deforestation from Amazonian rainforest by 2020, 77% coming from renewables by 2020, biofuel development and expansion of carbon markets (Cadena Monroy et al., 2011). There are also additional activities in agriculture, with activities embedded within its national plan, with the environmental subsector.

4.2. *Stage of development of action*

As described above, MAs are taken simply as those that mitigate GHG emission – whether having a climate objective or not. In arranging them, some underlying structure is being defined by all countries, even if not equally reflected in all MAs.

Brazil's MAs are at the most formal stage of development legally, encoded in a Law (12178). The implementation decree (7390) requires compiling further actions. An inter-ministerial committee consultation process with a multi-sector Brazilian Climate Change Forum was important in early stages of developing actions and is likely to remain active. The energy MAs have already been considered for a sectoral MA, and those in forestry and agriculture have specifically agreed goals. Ministries for different line functions have been required to submit plans in the first half of 2012.

Colombia in turn has advanced several sector-specific goals, with work on energy on incentives through dispatch and firm capacity credits. The forest certificates also have substantial potential, but will likely require regulatory changes to achieve goals. Colombia is probably the world's leading exporter in BRT systems, with BRTs already established in eight cities, with cutting edge design. Further actions may be required for scrappage and modal shift measures, and biofuel development also at forefront, but with constraints likely as a goal of 15% ethanol blend appears by 2020. Colombia is developing a LCDS as an overarching framework that will frame its MAs (Cadena Monroy et al., 2011).

Peru has a variety of actions in different stages. Their current suite of actions covers the energy, industry, transport, forestry and agriculture and waste sectors, with a specific lighting NAMA described (Takahashi et al., 2011). In the energy sector, these include renewable auctions, gas introduction for domestic use and transport, clean technologies in power generation, measures for smart driving, fleet and vehicle renewal and fuel improvements. Forestry also includes several regulatory and incentive instruments, while waste will see the construction of various landfills. The specific NAMA in lighting is currently being designed and analysed, together with a transport one. The intention is do further research as part of a stakeholder process, and then to move those less advanced into an implementation phase (Takahashi et al., 2011).

In contrast, Chile's and South Africa's are mostly in the design stage, with the MA in Chile mostly identified and now being designed in the energy and transport sectors, with preliminary work in forestry. In South Africa, most individual MAs are in design stage, with some advances on the BRT, a pilot in the settlement facility for housing. Tariff support would work within capacity licensing regime, and the tax support within the overall tax regime. South Africa's climate policy mandates the development of carbon budget for major sectors by 2013. MAs for each sector will need to be defined more clearly, as sectors (and some major entities) consider how to remain with the identified budgets.

4.3. *Institutional capacity*

The term 'institutions' is sometimes taken as the rules of the game, and in other sense it may mean agents and organization set up to enforce them. The term is used here in both senses, with differences noted as required. Institutional capacity is a key determinant for implementing MAs, and a country's mitigative capacity (Winkler, Baumert, Blanchard, Burch, & Robinson, 2007).

In all the cases, there are MAs linked to a country's policy and national objectives and goals, but more detailed MAs and actions seem to be linked to the enhanced enforcement, coordination and planning capacity. Several of these have considered how to articulate activities based both on the origins of the funding and the country's capacity to implement them. Likewise, existing planning and policy deployment capacity also seems to be a key to both the creation of MAs coordination and consultation, as well as their being implemented or not. In fact, the variation of capacity across sectoral institutions' is presented in the five country studies as a determinant of which are more likely to be implemented. This highlights that capacity development is a more significant matter than its relatively low status in negotiations might suggest.

The impact of institutional settings is felt in many different ways. In Brazil, MAs were collected and implemented by decentralized bodies. Implementation is in charge on national entities, with law enforcement and policy setting in Brazilian hands. Brazil has created new financial institutional capacity for its largest MA, the Amazon Fund to support reducing deforestation. The fund has received $1 billion of foreign funds and is housed within Brazil's Development Bank (BNDES), a major financial institution (see country study; La Rovere et al., 2011). The role of subnational governments is particularly relevant in deforestation and agriculture, with the subnational government, for instance, involved in policy implementation.

In Colombia, the country also has a well-organized business chamber structure, with coordinated research bodies and strong links with planning. Their ideas are included within national planning document and city planning documents; which are developed by central government, with some participation from development banks, policy research and input from private sector, civil society and private research. Institutions in the energy sector have enough capacity to design, implement and supervise MAs. For overall low-carbon development planning, there is significant capacity in the Environmental Ministry and Planning Department. There is also significant capacity in the forestry sector, but these require institutional learning to address the inherent difficulties in analysing and implementing MAs in deforestation. Colombia also has well-structured schemes in biofuels, regulated by the MoA, and with close links with regional research bodies in agriculture. A general feature of the institutional landscape in Colombia is that it is market-oriented, with regulation or planning needing to have a light touch in order not to be seen to distort markets. Another is a near-term focus, typically for 4 years of a Presidential term. This is a challenge in relation to climate change, due to its long-term nature, and for implementing responses to a market failure.

In Chile, the Ministry of Energy has dedicated agencies and expertise, which are not present in the Transport and Forestry sectors. However, the latter have long-standing technical expertise. Forestry (under Agriculture) has the capacity to define and design NAMAS, while transport can do so with the support of the Ministry of Environment. In Peru, there are variations on implementation within sectors, with energy the most comprehensive, including centralized bodies for planning, regulation and oversight, and with a current referential plan for efficiency. In forestry and transport, there are also several levels and agencies of government participating. This, together with high personnel rotation and low salaries conspire against stability and follow-up of proposals. Capacity is stronger in the energy sector, and less so in industry, transport and waste. Forestry had some overlap with Agriculture and Environment, creating some conflicts of jurisdiction. In South Africa, Capacity will need to grow at city government, to combine with that of central government.

4.4. *Planning and regulatory concept*

The strength of the planning and policy institutional setting and its regulatory environment affect the scope and character of MAs. All five countries have open market economies with regulatory and planning bodies set up to ensure market operation. Within this broadly similar context, there are different emphases. Some countries – for instance, South Africa or Brazil – have MA buttressed by institutional support within planning cycle, while others, such as Colombia, have these planning institutions, but rely for their implementation primarily on market institutions and interventions via price mechanisms – although these are indirect, they have the potential to powerfully affect MAs. Finally, others, such as Peru or Chile, rely primarily on markets and related regulation, rather than planning. Conversely, discussions on economic instruments can make a non-negligible effect on how planning is implemented – for example, South Africa's Treasury considers a carbon tax, which is taken into account in electricity planning.

In terms of how to bring consensus, Brazil has a permanent forum to consult actions, Peru has another committee, albeit more recent, with Colombia currently setting up one, while South Africa had extensive consultation in defining scenarios and then formal climate policy.

While all counties rely on regulation to some degree, Brazil, Colombia and South Africa seem to rely more on planning; Peru and Chile have had more of a free market-orientated structure.

From the articles, it is clear that there are, however, subtle but important differences among these planning bodies; as a result, planning does not mean the same thing in all countries. In Brazil, planning is done by ministries with input from technical bodies (in energy and forestry/agriculture), operating within sectoral legislation. Examples include energy auctions and tenders, forest code and credit eligibility requirements.

Colombia also has a strong planning and regulatory structure, with a planning department and regulatory commissions for specific issues. Information from all sectors is analysed by the National Planning Department (DNP), which coordinates across sectors and submits planning documents for the Presidential 4–5-year national plan. Given the mandate at the highest political level, the plan is widely respected. These plans run side by side with longer 10-year term prospective scenarios. DNP can also apply reliability charges within a free wholesale market – thus, planning is liked to the market incentive structure. In other areas, transport and BRT in turn have a close relation with local governments.

In the case of Chile, there is a free market operating in the energy sector, in contrast to several supervisory and regulatory mechanisms in forestry. There is no definition yet, however, on the composition and institutional standing of potential NAMAs within the forestry sector. Peru, like Chile, also has more of a pure market structure, with less of a planning structure, and a more indicative planning structure, advanced through CEPLAN (the Acronym for the National Planning Centre, in Spanish), under the purview of the Prime Minister of the President's Cabinet. While recently it is improving the capacity in the public sector, the country has in the last few decades relied extensively on the private sector to produce plans. In Climate Change, mitigation proposals included were generated by consultants using national mitigation guidelines, then consulted regionally and collected in national mitigation and adaptation plan.

4.5. *Technical capacity to design MAs*

In discussing these, the capacity to design MAs can be taken in two senses. The first would be to use existing government policy and sectoral planning capacity to design policy, which was addressed in the previous section. The second would be dedicated capacity, established specifically to address climate change.

An example of the latter from South Africa would be the capacity to design low-cost housing that provides in parallel lower emissions and also saves money for households and improves indoor air quality. Further work was then carried out to design a much larger MA, in the form of a facility that would replicate from the scale of a single Clean Development Mechanism (CDM) project to the scale of the whole housing programme. There is a detailed design of a National Sustainable Settlement Facility (G:enesis, 2008), but it is yet to be established. More generally, long-term coordinated action is affected by the prospective and planning horizons, the policy cycles and enforcement capacity and the stability of interaction among key agents.

Brazil probably has the most varied number of them: they are already designed, and are being implemented through national fora, with new MAs in course of being agreed and with direct and indirect policy mitigation instruments and vehicles.

There is also a strong record on flexible instrument implementation and regulation – interest in maintaining it. For example, in Chile and Colombia interest persists in the CDM, particularly relating to financing and ways to measure the aggregation of activities, as well as on financing. These created related MA activities. Chile, for instance, has created finance institutions that support investment with a mitigation impact, including through support by Chile's Production Development Corporation (CORFO, for its Spanish acronym), or aggregated activities around clean production.

Colombia has produced substantial mitigation innovation, for instance, in terms of Bus Rapid transit Systems, which affect both urban and mitigation policy, while there is on-going research in Forestry, biofuels and agriculture which is being translated into mitigation activities in those sectors.

In Peru, these are mostly currently developed from a coordinated mix of public, civil and private options through public private partnerships. These follow a strategic program approach with a rather innovative NAMA structure, combining carbon markets and self-financed action, ad developed in consultation with the private sector and called there a PRONAMI, from their Spanish acronym. The case for a lighting NAMA, also described in Takahashi et al. (2011) is another good example.

4.6. *Technical capacity to MRV, the implementation of actions*

Measuring, reporting and verifying (MRV) the implementation of nationally appropriate MAs is an important new requirement of transparency at the international level (UNFCCC, 2007, 2011). In spite of the fact that MAs without MRV are still happening independently of multilateral negotiations, MRV provides transparency internationally on how developing countries are implementing NAMAs. In this section, our primarily domestic focus on MAs is supplemented by this consideration – reflected also in the nomenclature of NAMAs as distinct from MAs.

From the studies by researchers in the five countries, it appears that the capacity for MRV begins with understanding on how to articulate a NAMA, and to attract international support. Technical capacity to MRV would relate to the availability, uniformity and character of data. There is greater experience and more certain data for energy-related NAMAs, with experience in the CDM having built technical capacity that is relevant to MRV. Even in this sector, there is great diversity of actions, with MRV of supply-side being directly measurable, whereas demand-side measures require a counter-factual. Likewise, the availability of Methodologies: the development of specific ones for transport, and the CDM were a good way of moving forward the process when BRT systems were established. Brazil in turn has established extensive measurement systems and data on deforestation. This is institutionalised in the National Institute for Space Research (INPE, in the Portugese acronym) and provides a solid technical capacity that could be applied to MRV of reducing emissions from deforestation and degradation (REDD) in Brazil and possibly other Amazonian countries.

Measurement is almost invariably done domestically, and reporting in the context of MRV is a communication by the country to the UNFCCC. Both are therefore firmly grounded in domestic activities. Crucial to transparency is the role of independent third-party data to verify what has been measured and reported. Technical capacity for verification would have an international component, but should also include the training of independent verifiers in developing countries.

Analysed from a sectoral perspective, the MRV in energy appears to be more straightforward, except for the choice of counterfactual for energy efficiency and demand-side management. The capacities identified in Sections 4.3, 4.4. and 4.5 for the five countries indicate that MRV for energy should be possible, although work remains to be done.

MRV in forestry would seem to be more complex. One issue is forest density, as satellite imagery provides good estimate of deforested areas. Such capacity is strong in Brazil, but less in Peru and Colombia. There is, however, a good potential for regional cooperation.

4.7. *Poverty and development*

In Brazil, there is a substantial use of credits in agricultural case, indirect in forestry, and with impacts on urban conditions, as well in the case of energy. In Colombia, there are strong implications in the deforestation and agricultural policies, and some indirect in transport, and some in energy and biofuels. Likewise, in Peru, there are strong implications in forestry and agriculture, indirect in energy and transport. In South Africa, there are strong and direct implications in the urban housing project, indirect in the others. In Chile, there are some impacts in energy and forestry.

Climate policy is about multiple objectives, both development and climate. Developmental objectives are themselves numerous, but across all five countries, alleviating poverty remains a high priority. Brazil and South Africa exhibit high levels of inequality. Climate is receiving more priority, but in the past it had not been at the top of the policy agenda in developing countries (Baumert & Winkler, 2005; Dubash & Bradley, 2005; La Rovere et al., 2007).

MAs that can show that they alleviate poverty, reduce inequality, contribute to socio-economic development and are much like those that gather broader societal and necessary political support (Rennkamp & Wlokas, 2012; Wlokas et al., 2012). While large segments of population remain in poverty, the key question will remain how to get out of it? This implies a discussion around development and changing from high- to low-carbon development pathways (Sathaye et al., 2009; Winkler & Marquard, 2009). It suggests that the concept of development needs to be redefined in relation to what it means to lead a good life and whether new concepts of growth and prosperity are required.

The link between mitigation and poverty poses analytical and methodological questions, as well as policy ones. A response is affected by various existing policies – notably not only agriculture and forestry but also food, water, housing, social policy, transport and energy. Moreover, such a response is solely not only an issue of trickling down growth but also of considering the role of dedicated policy: social and social inclusion policy, as well as that of deploying new forms of infrastructure, and lifestyles, and of preserving traditional and low carbon lifestyles and assets.

Can we imagine a future with radically new developments, or new combination between sectors rather than only derivation from those we have now? Are there overall development paths for countries that meet basic human needs (make poverty history) but with lower emissions than a fossil-intensive/high deforestation path? One way to address this is to forecast into the future based on existing trends and comparing results with back-casting from a future that meets developmental and poverty goals to the present. This would not be prescriptive – it is more an issue of how to fill in the blanks when imagining the future, and analysing multiple potential scenarios. Nevertheless, there are already several policies in place from where.

4.8. *Ownership*

Which actors or agencies conceptualise, design, plan and implement MAs? The 'ownership' of MAs will influence many aspects.

In Brazil, MAs are primarily initiated by public players, with private sector participation in implementation. In Colombia, government entities (national and subnational) are central, with some public–private partnership. In Chile, work on MAs was initiated by government, and strong government ownership is expected to continue, against possibly linking up in partnerships with private sector actors. In Peru, MAs are 'owned' by a mix of government and consultants, with some regional government participation. In South Africa, a mix of central and city government tends to develop on MAs, but there is also strong participation by civil society and private sector. The ownership of MAs is a fluid, rich, but still, yet to be coordinated environment.

4.9. *Finance*

In Colombia, there are implicit private sector incentives through firm capacity charges, additional financing options possible; in transport, multiple options operative with others in design (scrappage and retrofitting), plus incentives in forestry. There is potential additional action through international support. In Chile, in energy, there are some design issues around specific instruments (revolving funds, concessionary finance, subsidies and credits, etc.). Costs are estimated for forestry and energy, but there is not an agreed framework in the case of forestry yet. In Peru, there is strong potential for energy infrastructure activities to be privately financed; while in efficiency activities and those in other sectors, there might be a need to blending financial sources and carbon finance. Additional support for technical expertise might be required. Finally, in South Africa, there are several self-funded and innovative financial mechanisms (e.g. settlement facility and feed in tariffs) with clear overview of total costs, and some piloting experience. A particularly striking example of policy creating an enabling environment is the Brazilian Amazon Fund. The creation of the fund was made possible by the interaction between policy, legal implementation and the required resources to implement. International finance flows through BNDES and was blended with carbon markets and national support.

The consideration in the five studies on finance indicates no frequently blended single dominant source, but several. Various types of finance and financial agents participated in finding finance for MAs, including public, multilateral and private and domestic and international. In all countries, specific financial vehicles and implicit incentives were used (revolving funds secondary guarantees, private and public energy service company as agents etc.), as well as indirect and implicit incentives (capacity credits, interconnections, transmission, forest certificates, etc.).

5. A comparative table

Having described in some detail the concept of MAs in each of the five countries, stage of development, various capacities (institutional, regulatory/planning, technical design, MRV), issues of poverty and development, ownership and finance, the comparative analysis is synthesized in Table 1.

The table below compares the results. It uses the categories advanced above to organize information. Some remarks as to its character are worthwhile in this context. It is worth noting that MAs are not independent of the countries' resource base, and arise as an imaginative, political and economic response to it. Central in this is the development path and socio-economic characteristics influencing them. These can cover a whole suite of issues: gross domestic product (GDP), both in terms of growth rates and composition of GDP, levels of poverty and inequality, major economic sectors, levels of unemployment and size of population, predominant technologies, etc. Likewise, they are affected by the country resource endowments, emissions profile and scale. In terms of endowments, it is not the same to have abundant fossils than not to have them, or tropical forests than temperate ones. The country emissions profile in turn is derived from the use of the asset composition, while scale in turn affects emissions magnitudes, level of action and external pressures and opportunities. Finally, institutional factors affect agents and their interaction – their institutional development and stability – and are influenced in turn by policy cycles and political systems, and the underlying policy and national objectives.

Some additional associated, but less obvious, issues remain when addressing the above. Time is also of consequence: an early or late action has related costs – and affects where each MA is within a timeline and how much it can advance. While not the main focus, how to frame a MA as NAMA – and its associated MRV – also affects options, including, more generally, the relation between domestic and global action, and the relation of the former with long and very long-term visions of the country's development and, finally, the design, implementation and monitoring and review and implementation debate – the nature of the barriers being addressed affects the path taken.

6. Areas for further work

From the discussion above, there are several promising areas for further research and analysis. An obvious first area would be broadening the analysis, in particular to include other African and Asian countries. This study has been focused mainly on Latin America and in South Africa a somewhat atypical case. The findings across regional differences would be strengthened by an extended analysis.

This article has offered some initial reflections on the role that institutions and policy and planning horizons play in the definition and implementation of MAs. This could be deepened by reviewing literatures on institutional economics and public policy. These literatures may also offer further insights on the role of assets or resource endowments in determining which MA a country develops.

This article has focused on MAs and only briefly on transparency. The brief exploration of MRV could be taken further, including examination practice and cultural factors – how things are done in different countries. The role of independent verifiers in each country might be an illuminating example.

The cases also shed some initial light on the different origins of their MA and the different reasons behind them. In the case of Colombia, Chile and Peru, they decided to pledge a number of unconditional domestic actions as a way to both enhance competitive advantages and to elicit further ambition from the international community, thus reducing impacts (Garibaldi et al., 2012). Theirs was an early proactive link between mitigation and adaptation, in response to the developments within the international regime. Brazil and South Africa seem to have done the same later separately within the BASIC group, but focusing mostly on mitigation. As all these pledges were deployed, additional constraints and opportunities emerged, which had to be addressed, thus creating further developments. Comparing their different origins and how these affected their development path might be an area for further research.

In these developments, the various segments of the economy countries might face different challenges, with some focusing in reducing their emissions and yet others also in terms of avoiding future emissions. Likewise, how these fit within future visions of development and societies might become central. As time goes by, other developing countries might start using long-term pledges and then back-casting to the present. The comparison of scenarios developing through both fore- and back-casting may be an interesting area for future work.

Finally, the links between poverty and mitigation are perhaps the most interesting emerging new area, in our view, for future work. Such work should focus on both the relation between poverty and the development pathway, and on the potential for Mas to reduce poverty. Such work should also engage critically with concepts of growth and prosperity.

Overall, these areas could bring further insights on the reasons underlying why MAs get implemented or not, their policy and issue linkages across the economy and their diversity. Altogether, this promises to be a fruitful research programme.

7. Conclusion

These conclusions seek to expand some of the insights gained from the comparison against the objectives outlined in the introduction to this study.

Comparing the analysis by researchers from five developing countries makes it very clear that there is substantial MA on-going in Brazil, Chile, Colombia, Peru and South Africa. This in itself is an important first point.

MAs are driven by both responses to the international regime and the collective and individual climate impacts they are facing, as well as through their own domestic developmental and climate objectives. These actions' character, scope, policy horizon and potential success seem closely linked to resource base, institutional and policy settings, and the developmental path of the countries. MAs that address poverty and development – or alternatively, competitiveness and diversification concerns – appear to have a better chance of being implemented, since they address issues higher on the policy agenda in developing countries. MAs may develop effective through policy on energy, forestry, housing, transport, agriculture and through many other sectors.

Which policy linkages are established seems important in addressing sources of finance. The cases examined suggest that finance is typically a blend. The Brazilian case of creating a fund in its development bank, with a firm policy and legal basis, is particularly striking.

Some MAs are, however, also developed more specifically in response to the opportunities and requirements of the climate negotiations. In cases where MAs seek international support, it seems that all five countries have existing capacity that can be built upon to address MRV and transparency.

A similar finding relates to the institutional capacity to plan and regulate for MAs and to design them with technical competence. Which of the MAs are chosen relates to the emissions profile and resource endowment of countries. The case studies suggest that the implementation of MAs is undertaken sector by sector. Section 4.4 outlined the role of sector-specific approaches and instruments, including legislation in the case of Brazil, that seems important in these developing countries.

Moreover, the extent and strength of agents' capacity and coordination capacity across institutions might be a key reason why some MAs planned in these countries might or might not be implemented. Some countries, such as Brazil and Colombia, have both strong planning departments, side by side with Civil Society/Government fora. This allows to build up planning and legitimacy for their actions. Others have less planning bodies, or none at all, relying instead on the latter case – on markets. This situation expands from and also at city level. In several of the countries, public–private partnerships have emerged or are being envisaged.

Nevertheless, while this paper sheds some initial light on the different origins of MA, their drivers and how these were developed, there is more research needed on why these actions emerged.

In this context, the time horizons for policy seem to have affected the scope of the planning. Significant variation can be found, from 4-year plans to 40-year scenarios. Both perspectives are important, with the short-term being appropriate to the urgency of implementing MAs, but climate change requiring a long-term perspective. In fact, the evaluation of long-term scenarios, domestic, international and comparative, seems to have been at the root of the initial proposals advanced officially by Peru and informally by South Africa, at COP-14 in Poznan. In this, as in some other areas, these developing countries rather than following trends seem to have been leading the way.

References

Baumert, K., & Winkler, H. (2005). SD-PAMs and international climate agreements. Chapter 2. In R. Bradley & K. A. Baumert (Eds.), *Growing in the greenhouse: Protecting the climate by putting development first* (pp. 15–23). Washington, DC: World Resources Institute.

Cadena Monroy, Á. I., Rosales, R., Salazar, M., Rojas, A., Espinosa, M., & Delgado, R. (2011). *Mitigation options in Colombia* (Research report for MAPS (Mitigation Action Plans and Scenarios)). Bogotá, Colombia: Universidad de los Andes. Retrieved March 23, 2012, from http://www.mapsprogramme.org

Dubash, N., & Bradley, R. (2005). Pathways to rural electrification in India: Are national goals also an international opportunity? In R. Bradley, K. Baumert, & J. Pershing (Eds.), *Growing in the greenhouse: Protecting the climate by putting development first*. Washington, DC: World Resources Institute.

Garibaldi, J. A., Araya, M., & Edwards, G. (2012). *Shaping the Durban Platform: Latin America and the Caribbean in a future High Ambition Deal*. Policy Brief for Climate & Development Knowledge Network/Fundación Futuro Latinoamericano/Energeia. Retrieved from http://www.intercambioclimatico.com/wp-content/uploads/Shaping-Durban-Platform_Final-April_2012.pdf

G:enesis (G:enesis Analytics). (2008). *The national sustainable housing facility business and investment plan*. Cape Town: Genesis Analytics.

La Rovere, E. L., Dubeux, C., Perira, A., & Wills, W. (2011). *Mitigation actions in Brazil* (Research Report for MAPS (Mitigation Action Plans and Scenarios)). Rio de Janeiro, Brazil, Institute for Research and Postgraduate Studies of Engineering (COPPE) at the Federal University of Rio de Janeiro. Retrieved March 23, 2012, from http://www.mapsprogramme.org

La Rovere, E.L., Pereira, A.O., Simoes, A.F., Pereira, A.S., Garg, A., Halsnaes, K., ... da Costa, R.C. (2007). *Development First: Linking energy and emissions policies with sustainable development in Brazil*. ISBN: 978-87-550-3630-7. Roskilde: UNEP Risø Centre.

Ministerio del Ambiente. (2010). Plan de Acción de Adaptación y Mitigación frente al Cambio Climático. Retrieved from http://sinia.minam.gob.pe/

Rennkamp, B., & Wlokas, H. (2012). *A brief inspiration on poverty and climate change mitigation* (Policy brief for MAPS (Mitigation Action Plans and Scenarios)). Energy Research Centre, University of Cape Town. Retrieved from http://www.mapsprogramme.org

Republic of South Africa (RSA). (2008). Government's vision, strategic direction and framework for climate policy. Presentation by the Minister of Environmental Affairs and Tourism on the July Cabinet *lekgotla*. Cape Town: Department of Environmental Affairs and Tourism. Retrieved April 27, 2009, from http://www.environment.gov.za/NewsMedia/MedStat/2008Jul28_2/Media LTMS 29 July2008.ppt.

Republic of South Africa (RSA). (2010). Letter dated 29 January, for the South African national focal point. Pretoria: Department of Environmental Affairs. Retrieved February 8, 2010 from http://unfccc.int/files/meetings/application/pdf/southafricacphaccord_app2.pdf

Republic of South Africa (RSA). (2011). National climate change response white paper. Government Gazette No. 34695, Notice 757 of 2011. Pretoria: Department of Environmental Affairs. Retrieved October 26, 2011, from http://www.info.gov.za/view/DynamicAction?pageid=623&myID=315325 and http://www.environment.gov.za//PolLeg/WhitePapers/climatechange_whitepaper.htm

Sanhueza, E., & Palma, R. (2011). *Mitigation actions in Chile* (Research Report for MAPS (Mitigation Action Plans and Scenarios)). Santiago de Chile. Retrievied March 23, 2012, from http://www.mapsprogramme.org

Sathaye, J., Lecocq, F., Masanet, E., Najam, A., Schaeffer, R., Swart, R., & Winkler, H. (2009). Opportunities to change development pathways toward lower greenhouse gas emissions through energy efficiency. *Energy Efficiency, 2*(4), 317–337. Retrieved October 28, 2009, from http://www.springerlink.com/content/d71tg1r730m0674h/

Takahashi, T. P., Zevallos, P., & Cigaran, M. P. (2011). *Country study on mitigation actions in Peru* (Research report for MAPS (Mitigation Action Plans and Scenarios)). Lima, Peru, Libelula. Retrieved March 23, 2012, from http://www.mapsprogramme.org

Torres, M., Winkler, H., Tyler, E., Coetzee, K., & Boyd, A. (2012). *Mitigation Actions, NAMAs and LCDS: Building a common understanding*. Briefing Paper for MAPS (Mitigation Action Plans and Scenarios). Retrieved from http://www.mapsprogramme.org/

Tyler, E., Boyd, A., Coetzee, K., & Winkler, H. (2011). *Country study of South African mitigation actions* (Research report for MAPS (Mitigation Action Plans and Scenarios)). Cape Town: South Africa, Energy Research Centre, University of Cape Town. Retrieved March 23, 2012, from http://www.mapsprogramme.org.

UNEP. 2011. Low Carbon Development Strategies – A Primer on Framing NAMAs in Developing Countries. Retrieved from http://www.namapipeline.org/Publications/LowCarbonDevelopmentStrategies_NAMAprimer.pdf

UNFCCC (United Nations Framework Convention on Climate Change). (2007). Bali action plan. Decision 1/CP.13. Bali, Indonesia.

UNFCCC. (2011). Decision 2/CP.7: Outcome of the work of the ad hoc working group on long-term cooperative action under the convention.

Winkler, H., Baumert, K., Blanchard, O., Burch, S., & Robinson, J. (2007). What factors influence mitigative capacity? *Energy Policy, 35*(1), 692–703.

Winkler, H., & Marquard, A. (2009). Changing development paths: From an energy-intensive to low-carbon economy in South Africa. *Climate and Development, 1*(1), 47–65. Retrieved August 21, 2009, from http://www.ingentaconnect.com/content/earthscan/cdev/2009/.../art00006

Wlokas, H. L., Rennkamp, B., Torres Gunfaus, M., Winkler, H., Boyd, A., Tyler, E., & Fedorsky, C. (2012). *Low carbon development and poverty: Exploring poverty alleviating mitigation action in developing countries* (Research report for MAPS (Mitigation Action Plans and Scenarios)). Energy Research Centre, University of Cape Town. Retrieved May 4, 2012, from http://www.mapsprogramme.org/knowledge-sharing/research_lce-and-poverty-paper_120330/.

Index

Note: Page numbers in **bold** type refer to figures
Page numbers in *italic* type refer to tables
Page numbers followed by 'n' refer to notes

Printed and bound by CPI Group (UK) Ltd, Croydon, CR0 4YY
18/10/2024
01776204-0019